How to Repair
Briggs & Stratton
Engines

third edition

How to Repair Briggs & Stratton Engines

third edition

Paul Dempsey

TAB Books

Division of McGraw-Hill, Inc.

New York   San Francisco   Washington, D.C.   Auckland   Bogotá
Caracas   Lisbon   London   Madrid   Mexico City   Milan
Montreal   New Delhi   San Juan   Singapore
Sydney   Tokyo   Toronto

© 1994 by **TAB Books.**
TAB Books is a division of McGraw-Hill, Inc.

Printed in the United States of America. All rights reserved. The publisher
takes no responsibility for the use of any of the materials or methods
described in this book, nor for the products thereof.

The following are proprietary names of the Briggs & Stratton Corp. (listed
in order found in text): Vanguard, Rental Duty engines, Industrial Plus,
Europa, Cobalite, Rotocaps, Lube Power, Sprint, Magnetron, Magna-Matic,
Easy-Spin, Magnavac, Venturi, Vacu-Jet, Pulsa-Jet, Pulsa-Prime, Flo-Jets,
Quantum, System 3, System 4, and Kool Bore. Other product or brand
names used in this book may be trade names or trademarks. Where we
believe that there may be proprietary claims to such trade names or
trademarks, the name has been used with an initial capital or it has been
capitalized in the style used by the name claimant. Regardless of the
capitalization used, all such names have been used in an editorial
manner without any intent to convey endorsement of or other affiliation
with the name claimant. Neither the author nor the publisher intends to
express any judgment as to the validity or legal status of any such
proprietary claims.

pbk   7 8 9 FGR/FGR 9 9
hc   1 2 3 4 5 6 7 8 9 FGR/FGR 9 9 8 7 6 5 4

Library of Congress Cataloging-in-Publication Data

Dempsey, Paul.
    How to repair Briggs & Stratton engines / by Paul Dempsey. — 3rd
ed.
        p.   cm.
    Includes index.
    ISBN 0-07-016346-4   ISBN 0-07-016347-2 (pbk.)
    1. Internal combustion engines, Spark ignition—Maintenance and
repair.   I. Title.
TJ789.D43   1994
621.43'4—dc20   94-10827
                                                           CIP

Acquisitions Editor: April D. Nolan
Executive Editor: Joanne M. Slike
Managing Editor: Lori Flaherty
Editor: Barbara Minnich
Production Team: Katherine G. Brown, Director
                Susan E. Hansford, Coding
                Donna K. Harlacher, Coding
                Nancy K. Mickley, Proofreading
                Wanda S. Ditch, Desktop Operator
                Stephanie A. Myers, Computer Artist
                Joann Woy, Indexer
Design Team: Jaclyn J. Boone, Designer
                Brian Allison, Associate Designer
Cover design by Holberg Design, York, PA.
Cover photograph by Bender and Bender, Waldo, OH.   TAB3
Back cover copy written by Michael Crowner   0163472

0151052023

Contents

Introduction

This third revision of *How to Repair Briggs & Stratton Engines* is written for the millions of men and women who refuse to be intimidated by a lawnmower or a sump pump.

About three-quarters of the book consists of new material, including the first service information available to the public on the new-generation Europa engines. In addition, detailed service procedures are provided for all current and many older single-cylinder, side-valve engines. One new section covers the two-cycle engine widely used by professional groundskeepers.

A book of this type has to have a philosophy, a sort of guiding principle that determines what is included and how that material is presented. First, I wanted to stress troubleshooting because it is the hardest part of small engine repair. Once you know what the problem is, fixing it is easy. A whole chapter is devoted to this subject, beginning with why the engine won't start.

An "all-meat-and-no-potatoes" approach was taken to the writing, using the fewest possible words and the maximum number of illustrations. Mechanics need all the visual input they can get.

It also was important to make the book as comprehensive as possible, no matter how much the publisher objected to the additional pages. I do not believe that there has been a Briggs carburetor ever built that has been neglected and know of only one ignition system not covered. That system is so rare that most mechanics have never seen it.

The need to make the book comprehensive is also reflected in the many tables that are included. Most of you will never look at these tables but the information is there for those who need it. The tabular information also makes this book useful for professional mechanics who already know how to do the work but need hard reference data.

I also believe that engine repairs can be done economically. Far and wide, I am known as a skinflint. So, when a special factory tool is called for, I have included a picture of the tool as a guide for fabrication. Part numbers are also included, together with the addresses and phone numbers of all Briggs & Stratton regional distributors.

Finally, it seems necessary to briefly explain how the less obvious technology works. There is a satisfaction and an empowerment that comes from understanding things that, in the long run, might be of more value than just fixing engines.

Safety considerations

Hazards encountered in the normal course of repair/maintenance work are flagged in the text by **Warnings** and **Cautions**. **Warning** means risk of personal injury; **Caution** means risk of equipment damage. Only the most obvious hazards are called out. I cannot anticipate all the ways of getting hurt or of damaging engines. You should always read the repair steps in their entirety before you begin any procedure.

Safety depends on the attitude of the mechanic and, to no less a degree, on the environment in which he works. A disorganized, untidy shop invites accidents.

Do:

- Keep a fire extinguisher ready when servicing small engines.
- Handle gasoline with extreme caution. Refueling and any service operation that can result in spilled gasoline should be performed outdoors, away from possible ignition sources.
- Disconnect and restrain the spark-plug wire before working on the underside of rotary lawnmowers and other engine-driven equipment.
- Wear eye protection when using a grinder, power-driven wire brush, or hammer.

Do not:

- Refuel an engine while it is running or still hot. Allow at least five minutes for the engine to cool before refueling.
- Attempt to clear a flooded engine by removing the spark

plug and cranking. If compressed air is not available, install a fresh spark plug and crank with the throttle fully open.

- Operate an engine in an enclosed area. The exhaust contains carbon monoxide, a lethal gas.
- Operate an engine without a serviceable muffler or air filter. A spark arrester must be fitted for operation in wilderness areas.
- Tamper with governor settings, bend actuating links, or substitute governor springs. Excessive no-load speed might result in connecting-rod failure or flywheel explosion.

1

The product range

Briggs & Stratton is the largest manufacturer of small engines in the world and certainly the best known. Like Harley-Davidson (coincidentally also based in Milwaukee), Briggs & Stratton has become an American manufacturing legend by building simple, honest products that give value for the money. You can almost always trust a Briggs engine to run. When it fails, the cost of replacement parts is half or a third of what the competition charges. As of January 1992, a crankshaft for a 13-hp Honda GXV 390 cost $204.20; a shaft for a 14-hp Kohler CV14 cost $146.15. The same part for a Briggs 14-hp Vanguard 26170 listed at $93.90. A connecting rod for a Kawasaki FC 420V cost $37.52; the equivalent Briggs part cost less than $20 (data supplied by Briggs & Stratton).

There is also the matter of parts availability. Briggs engines require less parts inventory than most of the competition and, in the United States, are supported by 27 central sales and service distributors (Table 1-1). Parts and expert technical advice are close at hand.

Engine identification

Primary identification of a Briggs & Stratton engine is the five- or six-digit model number and a four-digit number, which are stamped directly on the cooling shroud or on a plate affixed to it.

Historically, the type number referred to mounting details, such as the dimensions and design of the power takeoff (PTO) end of the crankshaft. Mechanics could pretty much ignore it except when interchanging engines or crankshafts. However,

1

Table 1-1

Aiea (Honolulu)	HI	96701	Small Engine Clinic, Inc.	98-019 Kam Highway	808-488-0711
Ashland (Richmond)	VA	23005	Colonial Power Div. of RBI Corporation	101 Cedar Ridge Drive	804-550-2210
Billings	MT	59101	Original Equipment, Inc.	905 Second Avenue North	406-245-3081
Burlingame (San Francisco)	CA	94010	Pacific Western Power	1565 Adrian Road	415-692-3254
Carpinteria (Santa Barbara)	CA	93013	Power Equipment Company	1045 Cindy Lane	805-684-6657
Charlotte	NC	28206	AEA, Inc.	700 W. 28th Street	704-377-6991
Columbus	OH	43228	Central Power Systems	2555 International Street	614-876-3533
Dallas	TX	75207	Grayson Company, Inc.	1234 Motor Street	214-630-3272
Denver	CO	80223	Pacific Power Equipment Company	1441 W. Bayaud Avenue #4	303-744-7891
Elmhurst (Chicago)	IL	60126	Midwest Engine Warehouse	515 Romans Road	708-833-1200
Foxboro	MA	02035	Atlantic Power, Inc.	77 Green Street	508-543-6911
Houston	TX	77040	Engine Warehouse, Inc.	7415 Empire Central Drive	713-937-4000
Kenner (New Orleans)	LA	70062	Delta Power Equipment Company	755 E. Airline Highway	504-465-9222
Louisville	KY	40299	Commonwealth Engine, Inc.	11421 Electron Drive	502-267-7883
Memphis	TN	38116	Automotive Electric Corporation	3250 Millbranch Road	901-345-0300
Milwaukee	WI	53209	Wisconsin Magneto, Inc.	4727 N. Teutonia Avenue	414-445-2800
Minneapolis	MN	55432	Wisconsin Magneto, Inc.	8010 Ranchers Road	612-780-5585
Norcross (Atlanta)	GA	30093	Sedco, Inc.	4305 Steve Reynolds Blvd.	404-925-4706
Oklahoma City	OK	73108	Engine Warehouse, Inc.	4200 Highline Blvd.	405-946-7800
Omaha	NE	68127	Midwest Engine Warehouse of Omaha	7706 "I" Plaza	402-339-4700
Phoenix	AZ	85009	Power Equipment Company	#7 North 43rd Avenue	602-272-3936
Pittsburgh	PA	15233	Three Rivers Engine Distributors	1411 Beaver Avenue	412-321-4111
St. Louis	MO	63103	Diamond Engine Sales	3134 Washington Avenue	314-652-2202
Salt Lake City	UT	84101	Frank Edwards Company	1284 South 500 West	801-972-0128
Somerset (New Brunswick)	NJ	08873	Atlantic Power, Inc.	650 Howard Avenue	908-356-8400
Tampa	FL	33606	Spencer Engine, Inc.	1114 W. Cass Street	813-253-6035
Tualatin (Portland)	OR	97062	Brown & Wiser, Inc.	9991 S. W. Avery Street	503-692-0330
CANADA					
Mississauga (Toronto)	ON	L5T 2J3	Briggs & Stratton Canada Inc.	301 Ambassador Drive	416-795-2632
Delta (Vancouver)	BC	V3M 6K2	Briggs & Stratton Canada Inc.	1360 Cliveden Avenue	604-520-1294

as engines become more complex, the type number has become increasingly critical. For example, Model 99700 lawn-mower engines with 3000-series type numbers have flywheel brakes.

The production mix has become so complex that Briggs recently added two digits to the end of the type number, a code number indicating the build date and factory. For example, 93011510 translates as follows:

> 93 = 1993
> 01 = January
> 15 = 15th day of the month
> 10 = plant

In some instances, the build date must be supplied to obtain the correct replacement parts. The model number code follows this rough logic:

initial one or two digits	cubic inch displacement*
1st digit after displacement	design series, e.g., I/C
2nd digit after displacement	crankshaft orientation (vertical or horizontal), carburetor and governor type
3rd digit after displacement	main bearing type, reduction gears, auxiliary PTOs, oil pump
final digit	starter type (notched pulley, rope rewind, electric, etc.).

* CID is an approximate number, usually rounded up to nearest cubic inch.

Modern Briggs & Stratton engines fall into three product lines: Standard series, industrial/commercial (I/C), and Vanguard. This book covers service procedures for single-cylinder, side-valve Standard and I/C models. The Vanguard series has little generic relationship with other Briggs engines, although carburetors, starters, charging systems, and incidental components are similar.

Standard series

Standard series engines are derived, for the most part, from those built in the 1960s and earlier. With one exception, all have side valves, which are actuated by a camshaft in the crankcase. Lubrication is by splash. Pistons generally run against chrome-plated aluminum bores.

Some things have changed. Overall quality is far better than it used to be. Magnetos have been replaced by solid-state pulse generators, and paper air filters, sometimes with a precleaner, have appeared on certain models.

The Standard series also includes the vertical crankshaft Europa. With overhead valves, cast-iron bore liner, and semi-pressurized lubrication, the Europa is unlike anything Briggs has ever built. It develops 5 hp for 9 cubic inches of displacement, or about 30 percent more power than a comparable side-valve model.

Briggs keeps inventing new models, most of them in the Standard series. Many of these exercises are cosmetic, such as, the four levels of trim currently offered for vertical shaft models. One can order an XE, XTE, XTL, or XM, each distinguished by the shape of the gasoline tank and the plastic cover over the blower shroud. The Max, Diamond Plus, and especially the Quantum represent more substantive changes that extend to the hardware and camshaft timing, and affect operating characteristics. One version of the Quantum makes a full hp more than the model from which it was derived. High-volume customers can style and, within limits, design their own engines, thanks to flexible manufacturing techniques. The real story of Briggs & Stratton is the way they run their factories.

I/C series

The manufacturer describes the I/C series as "mid-market" power plants, suitable for medium-duty applications. The series derives from standard engines, but in ways that are complex and difficult to generalize. One feature that all I/C engines have is an iron cylinder bore, either by virtue of an iron block or an iron liner in an aluminum block. Iron cylinders probably live longer than the chromium plate and can be rebored when worn. Three all-iron engines, now built overseas, are part of the line.

I/C engines tend to become more sophisticated as they get larger and more expensive. The best of them use high-temperature valves, two-stage air cleaners, centrifugal governors, and float-type carburetors. In a few instances, the I/C version develops more power.

The series also includes sub-variants. The Industrial Plus 5-hp engine appears identical to the I/C version, except that it features a larger fuel tank and a float-type carburetor, rather than the old-fashioned Pulsa-Jet. Unaccountably, the Industrial Plus lists for about $10 less than the I/C. Rental Duty engines, available in 3, 5, 7, 8, 10, and 11 hp, with a horizontal crankshaft, are more expensive versions of equivalent I/C engines. They are painted yellow and fitted with what the factory calls a "graphics package" and muffler guard. Fuel tank capacities, crankshaft configuration, and other technical details might also differ.

Readers who are familiar with small engine basics can skip the two following sections. Those who are new to the subject should read the material carefully. Little that follows will make much sense unless you know the names of the parts and have a fundamental understanding of how engines convert air and gasoline vapor into mechanical work.

Nomenclature & construction

Figure 1-1 identifies the major components that make up a typical single-cylinder, air-cooled, side-valve engine. The example illustrated uses ball-type main bearings, splash lubrication, and a centrifugal (flyweight) governor. These features are fairly unique to small engines.

Air cooling

The flywheel doubles as a centrifugal fan. Impeller blades on the rim draw air from the central hub area and discharge it into a sheetmetal shroud. Fins, cast into the cylinder head and barrel, transfer about a third of the heat from combustion to the atmosphere.

Valves

The valves on most small engines live in the block, alongside the piston, as shown in the referenced drawing. While this arrangement makes for mechanical simplicity, the awkward shape of the combustion chamber limits compression and probably encourages emissions. The Europa is the first of a new generation of overhead-valve (OHV) utility engines.

Premium engines use exhaust valves made of a cobalt alloy, which Briggs & Stratton calls Cobalite. Some models also feature positive valve rotation (via Rotocaps).

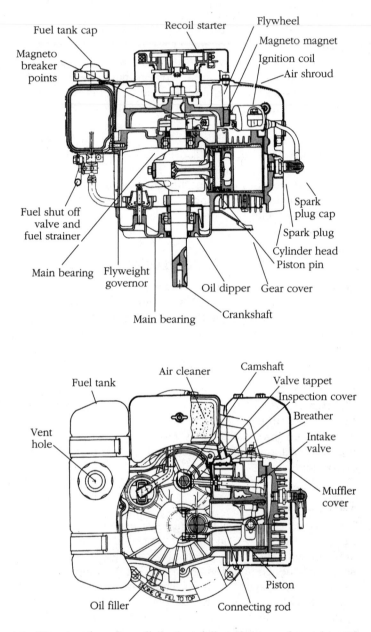

1-1 *Mass-produced small engines follow the same formula, with only subtle differences between the most and least expensive Wisconsin Robin W1-145V shown.*

Crankshaft bearings

Ball bearing mains are a traditional sign of quality in small engines, although well maintained plain bearings last about as long and run quieter in the bargain. Some I/C and standard engines split the difference, using plain bearings (or replaceable bushings) on the magneto end of the crankshaft and balls on the PTO end, where side loads are greatest.

Only a handful of four-cycle engines, none of them manufactured by Briggs, use insert-type connecting-rod bearings. When the rod wears, it must be replaced, usually in conjunction with the crankshaft.

Governor

Small utility and industrial engines include a governor, which limits no-load speed and automatically adjusts engine speed to changes in load. Mechanical governors, such as the one shown in Fig. 1-1, generally are more responsive than pneumatic governors, which sense engine speed as the dynamic head of cooling air against a spring-loaded vane.

Lubrication

Side-valve engines splash oil about the crankcase with a paddle wheel that Briggs calls a "Lube Power gear-driven lubrication system." The Europa supplements the paddle with a small pump.

Operation

Most internal combustion engines operate on a four-part cycle, consisting of the following events:

- intake of fuel vapor and air through a valve or port;
- compression of the fuel charge by the piston;
- ignition and subsequent pressure rise in the cylinder; and
- exhaust of the spent gases.

Four-stroke-cycle engines require four piston strokes, or two complete revolutions of the crankshaft, to complete the cycle. Two-stroke-cycle engines telescope events into two piston strokes, or one crankshaft revolution.

Four-cycle engines

Figure 1-2 illustrates the sequence of piston and valve movement during the four-stroke cycle. The cycle begins with the

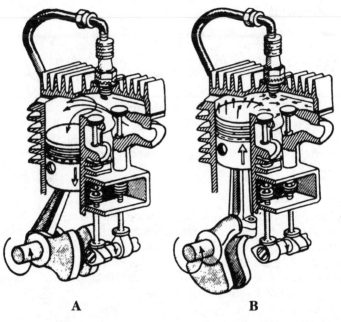

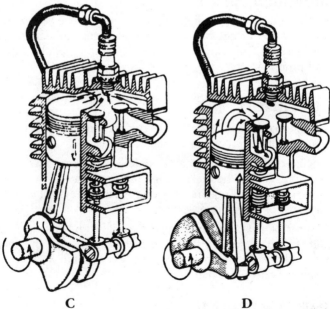

1-2 *The four-stroke cycle: A) intake; B) compression; C) power, or expansion; D) exhaust.* Tecumseh Products Co.

piston moving downward on the intake stroke. Air and fuel, impelled by atmospheric pressure, enter the cylinder at the intake valve. The exhaust valve remains closed during the intake and subsequent compression stroke.

As the piston reaches the lower limit of its travel, known as bottom dead center (bdc), the intake valve closes to seal the cylinder. The piston begins to move upward in the compression stroke. Air and fuel are compressed between the top of the piston and the underside of the cylinder head.

As the piston approaches top dead center (tdc), the spark plug fires, igniting the charge. Impelled by the force of the explosion, the piston dives toward bdc in the power stroke. The flywheel accelerates, absorbing energy that will be returned to the system during the exhaust, intake, and compression strokes.

The exhaust valve opens as the piston rounds bdc. Spent gases spill into the atmosphere, initially blown down by the residual pressure in the cylinder and finally by the piston. The exhaust valve closes near tdc and the intake valve opens to initiate another cycle.

Two-cycle engines

Briggs markets a pair of two-cycle engines. The horizontal shaft 62032-0529 is the least expensive of the lot and can be found on inexpensive snow blowers. It is not discussed here. The 4-hp I/C 96700 is a more serious device. It is marketed to professional landscapers who like its flat torque curve and immunity to damage if you fail to check the crankcase oil.

The piston is the central element of a two-cycle engine. It functions both as a double-action compressor and, in conjunction with ports in the cylinder bore, as a shuttle valve. The mixture is compressed in the cylinder bore above the piston and in the crankcase below it. Moving down the bore from tdc, the piston first uncovers the exhaust port then, somewhat later in the cycle, uncovers the transfer port that connects the crankcase with the upper-cylinder bore. A third piston-controlled port opens near the top of the stroke to admit air and fuel into the crankcase. Because the crankcase is part of the induction tract, it contains no oil. Lubrication is accomplished by mixing oil with the fuel. Two-cycle engines are mechanically simple and geometrically complex.

Intake In Fig. 1-3A, the piston approaches bdc. Impelled by the descending piston, the air/fuel/oil mixture in the crankcase undergoes a slight pressurization, sufficient to force it through the open transfer port and into the cylinder above the piston. The fresh charge enters with such velocity that it displaces (scavenges) the exhaust gases that remain in the cylinder, forcing them out through the open exhaust port.

Compression   In Fig. 1-3B, the piston has rounded bdc and is rising. Because both the transfer and exhaust ports are closed, the fuel mixture comes under increased compression. The ascending piston also depressurizes the crankcase cavity and uncovers the third port. The partial vacuum causes a fresh charge to flow from the carburetor into the crankcase.

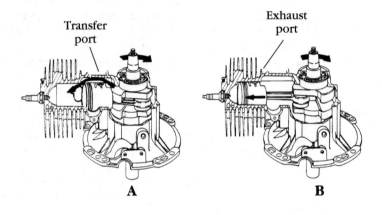

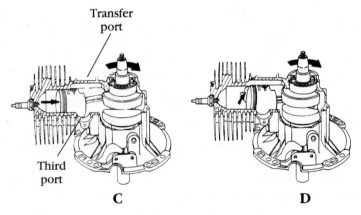

1-3   *The two-stroke cycle: A) intake; B) compression; C) power or expansion; D) exhaust.*

Power   The spark plug fires and ignites the fuel charge, as the piston rounds tdc. The power stroke (Fig. 1-3C) begins.

Exhaust   The power stroke continues until the exhaust port opens to blow down cylinder pressure (Fig. 1-3D). By the time this occurs, the piston has already moved down far enough in the bore to cover the third port. The crankcase feels increasing pressure, which peaks just before the transfer port opens to initiate a new cycle.

At working speeds, usually from about 2000 rpm to the governed limit, everything goes smoothly: gas flows synchronize and the engine seems to grow an extra cylinder. At lower speeds and under light loads, fuel escapes from the exhaust port and spits back through the third port. If you were to operate the engine without an air cleaner—a practice emphatically not recommended—you could see the haze of oily fog standing in front of the air horn. Flow dislocations can cause "four stroking" at idle, as the engine misses and hits with a bang on the next revolution.

Displacement, hp, torque, and rpm

It is no accident that the first set of digits in the Briggs & Stratton model code express displacement in cubic inches. Displacement—the cylinder volume swept by the piston—is the ground-zero measurement of engine potential, in the same way that square footage is the single most important variable in the design of a house.

bore × bore × number of cylinders × stroke × 0.7858 = displacement

If bore and stroke are expressed in inches, the formula gives displacement in cubic inches (in.^{3}); if in millimeters, displacement comes out as cubic millimeters (mm^3). Dividing the answer by 1000 converts to cubic centimeters (cc), the international standard for cylinder displacement. One cc equals 16.39 in.^{3}.

The 3.75-hp Sprint Model 96900 has a 2.56-in. (65.0-mm) bore and a 1.75-in. (44.4-mm) stroke.

2.56 in. × 2.56 in. × 1 × 1.75 in. × 0.7858 = 9.01 CID

Performing the same calculation with metric units:

65.1 mm × 65.1 mm × 1 × 44.4 mm × 0.7858 = 147332.8 mm^3 ÷ 1000 = 147.3 cc

Hp expresses the rate of doing mechanical work—pumping water, cutting grass, generating electrical power, or whatever. Hp seems like a straightforward concept and has sold many engines since James Watt coined the term in the 18th Century. But the horses promised by the manufacturer might not come when you summon them.

Power depends on many factors, such as displacement, compression ratio, volumetric efficiency, and engine speed. The kicker is rpm. Figure 1-4A shows the power curves for two Briggs & Stratton engines of almost identical displacements. One is a carryover from the days when engines were made of iron and develops 10 hp; the other represents a latter generation and produces a nominal 11 hp.

The 11-hp engine is more powerful than the iron antique, but that advantage is limited to the upper-end of the rpm scale. At 2600 rpm both engines produce slightly more than 8 hp. Drop the speed to 2000 rpm and the older engine begins to show the advantages of conservative valve timing. At 1800 rpm there is no contest.

In other words, the shape of the hp curve is at least as significant as rated hp. Hard-used utility engines need power at high rpm. Industrial engines that slog along at part throttle are better served by flat power curves.

Torque is a measure of instantaneous twisting force (in pounds or Newtons) on an imaginary beam, the length of which is expressed in feet or meters. Horsepower is what a man exerts when he rides a bicycle at a steady speed on a level surface; torque is the heave required to break a stubborn bolt loose.

The ideal engine would have a flat torque curve throughout the rpm range, but that is difficult to arrange. Torque peaks around three-quarters throttle. What happens on either side of the peak determines the flexibility of the engine, and its speed sensitivity to load.

The iron engine builds torque like a locomotive, up to 2600 rpm peak and then drops off sharply (Fig. 1-4A). The newer model doesn't do as well at low speeds but holds its peak in an almost horizontal line to the rpm limit. This engine is obviously intended to operate at wide throttle angles.

The shape of the torque curve is influenced by valve diameter, port length and shape, and by the camshaft grind. A small-valved engine with restrictive porting and a conservative camshaft tends to make torque early and then lose it at high rpms. An engine that flows better is happiest at higher speeds.

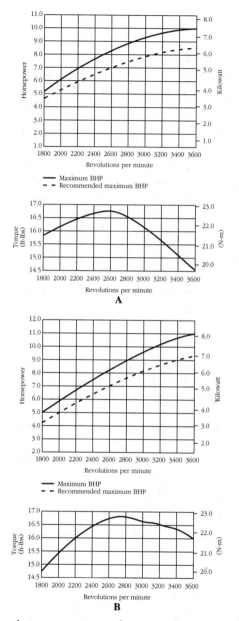

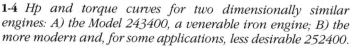

1-4 *Hp and torque curves for two dimensionally similar engines: A) the Model 243400, a venerable iron engine; B) the more modern and, for some applications, less desirable 252400.*

High levels of torque at wide throttle angles also influences the hp curve, skewing it to the left.

Purchasing an engine

Your area distributor (see Table 1-1) should be able to provide you with a copy of the current Briggs & Stratton *Engine Sales Replacement Specifications and Price List*, which contains the performance, pricing, and dimensional data necessary for intelligent shopping. This document does not supply hp curves for all models but it does show torque curves, which are a way of cutting through the complexity of marketing names. For most applications, you will not go wrong if you select the engine with the flattest, highest torque curve in its displacement class.

The critical dimensions for any application are the crankshaft PTO-side stub diameter, extension, and detailing. For example, the horizontal 7-hp Model 170400 can be obtained (under various series numbers) with any of five crankshafts. The same is true of vertical shaft engines. The direction of gear-type PTOs have a 6:1 reduction ratio, but rotation varies, as indicated by the series number.

Critical dimensions for horizontal shaft engines are the vertical distance from the centerline of the crankshaft-PTO stub to the engine mounting surface and the mounting bolt location. All horizontal shaft engines have a rectangular footprint with one mounting bolt in each corner. Naturally, both crankshaft height and footprint size tend to become larger with engine displacement. You can sometimes redrill mounting holes on the driven equipment, but variations in crankshaft height might be more difficult to accommodate.

The footprint of vertical shaft engines—the overall diameter of the flange, the size of the mounting hole, and the radius of the mounting bolt circle—varies with displacement. It is, however, often possible to substitute a slightly larger and more powerful engine for the one supplied without getting into trouble, but check the drawings first. Note that certain engines designed for Toro and Jacobsen mowing machinery have special flange castings.

The choice between Standard and I/C engines probably hinges on the intended use and the value you ascribe to the factory warranty. In many instances, the I/C appears to be a bargain, however, you might question the wisdom of purchasing a 4-hp I/C 114902 at the suggested list of $241.70 when the Standard 110902 lists for $180.20. The I/C engine has an iron bore, a mechanical governor (rather than an air vane), a less mainte-nance-intensive air filter, and a larger muffler with a wire mesh guard around it. These features add one-third to the cost of the base engine. However, this and all other I/C products carry a two-year consumer warranty and a one-year commercial warranty. Standard engines come with a one-year consumer warranty only.

Finally, there is the question of price structure. If you walk into a Briggs & Stratton dealer, you will pay the suggested list price and often a surcharge for "freight." Granger and other high-volume industrial supply houses have better prices but limited selection. If you need a basic vertical shaft engine, you might come out ahead by buying the engine as part of an inex-pensive lawnmower.

Dealers and distributors will sometimes negotiate on prices for individual engines, but don't expect miracles. If you intend to make a major purchase involving several engines, it is good business to mail a bid-request form to area distributors and dealers. Include model and serial numbers, quantity, your cut-off date for accepting bids, and terms of payment.

2

Troubleshooting

An engine will run as long as it has spark, fuel, and compression. An engine will start if the flywheel is spun rapidly enough to generate spark voltage and draw fuel into the cylinder.

Four distinct systems are involved:

- ignition
- fuel
- starting
- engine mechanical (consisting of internal crankcase and cylinder parts)

This chapter provides a structured approach to the sometimes difficult task of relating engine malfunctions to one or another of the four basic systems. Once the problem is identified, you can turn to the appropriate chapter for additional troubleshooting information and detailed repair instructions.

Before you begin

Troubleshooting is, in part, an intuitive process that depends on all sorts of subliminal clues. You need to spend a few minutes "smoozing" in the same way that good doctors are never too busy to engage their patients in small talk. While you are at it, change the consumables—the oil, fuel, air filter, and spark plug. Try to get a sense of the engine's history, the way it was used and maintained. It is always helpful to talk with the operator.

Four-cycle lubrication

The oil level should register full on the dipstick or reach the top thread of the filler plug port on four-cycle engines. Black, carbonized oil that feels gritty when rubbed between the fingers is a sure sign of trouble. The engine might not be worth the cost of repair, since all bearing surfaces will be damaged, with crankshaft and flange bearing (on vertical-shaft engines) getting the worst of it.

The factory specifies detergent oils labeled "For Service SE, SF, FG." Viscosity recommendations vary with ambient temperature and engine type (Table 2-1).

Table 2-1.   Viscosity recommendations

Engine type	Oil grade	Ambient temperature anticipated before next oil change (°F)
Side valve	Synthetic 5W-20, 5W-30	−20° to 40°
	Conventional 5W-30, 10W-30	0° to 40°
	Conventional 30	40° and above
Overhead valve	Conventional 5W-20, 5W-30	−20° to 20°
	Conventional 10W-30	0° to 100°
	Conventional 30	40° and above
Two cycle	TCW-30W, 1:50 ratio with fuel	Winter and summer

Two-cycle lubrication

Because it is impossible to qualify the amount or type of oil in the fuel, drain the tank and start with a fresh mixture. Briggs & Stratton recommends a 50:1 gasoline/oil ratio for the Model 96700 two-cycle engine. Working outdoors and away from possible ignition sources, mix 1.25 ounces of BIA or NMMA-certified TC-W30-weight oil with 0.5 gallons of unleaded gasoline (not gasohol) in a separate container. Shake vigorously and refill the tank.

Warning: Handle gasoline outdoors in an area away from potential ignition sources. Wipe up any spills and allow ample time for the residue to evaporate.

Fuel

Gasoline has a shelf life of less than six months, before the light hydrocarbons (which make cold starting possible) evaporate. Unless you are certain of its quality, drain and discard the contents of the tank. If the fuel smells acrid or is discolored with rust particles, clean the tank and carburetor, and replace all "soft" parts in the fuel system. Gaskets and diaphragms will almost certainly have been damaged.

The factory recommends unleaded gasoline, purchased in small quantities, and used within 30 days of purchase. Gasohol, most of which contains 10 percent ethanol, can be tolerated if drained from the tank and carburetor during storage. Gasohols with a higher percentage of ethanol or those made with methanol are quite corrosive and not recommended. In practice, this means that you should not use any gasohol because pump labels cannot be trusted.

Note: Winter grades of gasoline pose a special hazard for small engines because of their close-coupled fuel and exhaust systems. For easy starting, winter gasolines are formulated to auto-ignite at temperatures in the 400-degree-Fahrenheit range.

Warning: Handle gasoline outdoors away from potential ignition sources. Wipe up any spills and allow ample time for the residue to evaporate before you attempt to start the engine or test the ignition output.

Air filter

A dirty air filter element acts like a choke and enriches the mixture. In the short-run, a restricted filter will accelerate upper-cylinder bore wear by diluting the oil. Over the long-run, paper elements and sealing gaskets will give way and restore the mixture to its original calibration, but will allow abrasives to enter the engine. Clean and re-oil polyurethane (sponge) elements as described in chapter 4. Replace pleated paper elements as a matter of course.

Spark plug

Replace the spark plug with a known good one of the correct heat range and type. Once or twice in a working lifetime a mechanic encounters a new spark plug that refuses to fire in small engines. Although the possibility of this happening is extremely remote, it is not a bad idea to test spark plugs in a running engine before you use them as diagnostic tools.

Note: Verify that the spark-plug lead makes positive contact with the spark-plug terminal.

Begin by attempting to reproduce the mode of failure. The work you have already done might have solved the problem. If it doesn't, you will at least have the satisfaction of knowing that the complaint is real.

Testing the ignition output

The first step to testing the ignition output is to remove the spark plug and test the output voltage on all engines, whatever the complaint. The ignition system is the primary suspect, especially on older, magneto-fired engines.

No spark	engine refuses to start or stops when malfunction appears
Insufficient voltage	difficult starting, misfiring under sudden throttle or heavy load
Erratic spark	ragged idle, loss of power, and possibly "hunting" as the engine drops and regains rpm

Any ignition malfunction lowers the temperature of the spark plug. A no-spark condition tends to flood the cylinder and plug with raw fuel. A weak or erratic spark carbons over the firing tip—exactly as if the mixture were too rich.

Test spark output with PN 19051, available from Briggs & Stratton regional distributors for about $5. The traditional method of checking spark output by inserting a screwdriver into the spark-plug cable terminal is not recommended, primarily because it creates an ignition source outside of the engine. The Briggs tool isolates the spark safely behind a window.

Connect the tool as shown in Fig. 2-1A, with the larger (0.166-in.) gap between the cable terminal and engine ground. Remove the spark plug and spin the flywheel rapidly (at least 350 rpm). Watch for a spark. Newer engines use solid-state Magnetron spark generators, which deliver reddish, spindly sparks touched with blue. That spark should be sufficient, so long as it is regular. Older models were equipped with magnetos of various types. These systems should deliver fat, electric blue sparks, hot enough to burn paint. A white or reddish Magnetronlike spark from a magneto means trouble.

A spark miss can be detected by connecting PN 19051 in series with the spark plug and running the engine (Fig. 2-1B). Use

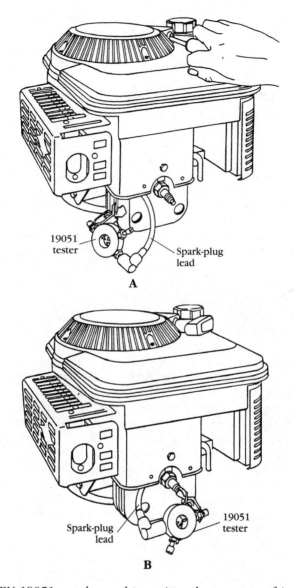

2-1 *PN 19051 can be used to register the presence of ignition voltage during cranking (A) and to detect voltage interruptions in a running engine. (B) In neither case does the tool say anything about spark-plug performance, which can only be determined by substitution of a known good plug. Also note that a spark gap in series with the spark plug (B) helps clear flooded engines by boosting system voltage.*

the wider (0.166-in.) gap for all modern engines. The narrower (0.060-in.) gap is for testing low-voltage Magna-Matic magnetos found on ancient models 9, 14, 19, and 23 and on very early 191000 and 23100 engines.

Replacing the flywheel key

Regardless of spark quality, the next step is to replace the flywheel key following the instructions in chapter 3. The key locates the flywheel relative to the crankshaft. Any misalignment affects timing and might reduce ignition voltage. These faults might not show on a spark-gap test, especially when a Magnetron system is involved. If spark quality remains poor after the key has been changed, locate and repair the fault following the instructions given in chapter 3.

At the risk of boring you with what must be obvious to most of you, it should be said that satisfactory ignition output and flywheel key renewal form the baseline for all that follows.

Starter output & compression

Testing the ignition system also tests the starter and provides some notion of engine compression. In general, it is easier and more informative to crank engines the old-fashioned way—by hand. If you use an electric starter, charge the battery, clean the cable terminals, and note that starter motors have a one minute duty cycle, followed by a 15 minute cool down.

Engage the rewind or electric starter. If the rewind starter cord binds or if the electric starter bogs, stalls, or runs at uneven speed, remove the cooling shroud and attempt to turn the flywheel through a full revolution by hand.

Warning: Disconnect and secure the spark-plug wire to prevent accidental starting. Disconnect the positive battery cable on electric-start models.

You should feel compression build and a varied but smooth resistance as the piston moves through its operating cycle. Resistance will increase slightly at mid-stroke, where piston velocity and friction are greatest. If the flywheel binds or encounters "hard spots," disassemble the engine to discover why. If the engine turns over normally, the problem is in the starter or starter alignment.

Small engine mechanics rarely attempt to measure cylinder compression with a gauge. Briggs & Stratton and other engine manufacturers have made it too difficult to do by incorporating automatic compression releases in most four-cycle engines. The most commonly encountered Briggs device, known as Easy-Spin, incorporates a ramp on the intake cam lobe that raises the valve slightly off its seat.

This and starter-engaged compression releases can be defeated by removing the shroud and spinning the flywheel against the direction of normal rotation. The flywheel should rebound smartly, as if from contact with a spring. But any indication of compression, as during normal cranking, should suffice.

Warning: Avoid contact with the sharp edges of the governor air vane, present on most of these engines.

There also is no easy way to measure two-cycle crankcase compression, but an experienced mechanic can sense its presence by the resistance on the starter cord. As a point of interest, crankcase compression typically amounts to 3 to 4 psi.

Testing the fuel system

Once you are satisfied with the spark and the integrity of the flywheel key, turn your attention to the fuel system. If you have done the preliminary work, you have cleaned or replaced the air filter element and put fresh gasoline in the tank.

Without the tools normally available to a small-engine mechanic, we cannot directly verify that the proper air/fuel mixture goes to the engine. But we can "read" the spark plug for indirect evidence of mixture strength.

Engine does not start

If the tip remains dry after prolonged cranking, you can be sure that no fuel is reaching the cylinder. Verify this by spraying a small amount of Wynn's Carburetor Cleaner or an equivalent product into the spark-plug port. Replace the spark plug and try to start the engine:

Engine starts on carburetor cleaner and continues to run This type of behavior indicates a weak starting mixture, probably caused by an improperly adjusted choke. The choke valve must close fully—completely blocking the carburetor bore—for most engines to start (Fig. 2-2). Primer pumps, used on nylon-bodied, suction-lift carburetors in lieu of a choke, must be cycled at least three times for fuel to enter the carburetor bore. If the bore remains dry after repeated pumping, you can assume that the primer bulb or check valve has failed.

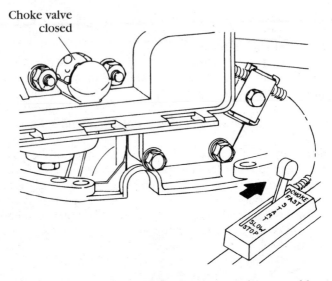

Choke valve
closed

2-2 *The choke, whether automatic or manual-remote (shown), must close fully for reliable starting.*

Engine runs a few seconds and dies   The cause is fuel starvation. Check the fuel system as detailed in chapter 4 to find the fuel blockage upstream of the throttle valve or vacuum leak downstream of it.

If the spark-plug tip is wet with raw fuel, the cylinder is flooded. It is normal for the spark-plug tip to grow damp with fuel in a nonstarting engine. And it is normal for the spark plug to become progressively wetter as cranking continues. In other words, any recalcitrant engine with a functional fuel system will flood if you crank long enough.

The engine can also flood of its own accord because of a failed float or inlet needle and seat. The carburetor bore will be awash in fuel. Suction-lift carburetors, which mount on top of the fuel tank, are immune to this type of failure.

Also note that tilting a four-cycle engine with its cylinder head down or prolonged cranking can splatter oil over the combustion chamber. In addition to being very difficult to clear, oil-flooding can mislead you into thinking that the engine is getting fuel when it might not be. The spark-plug tip should smell strongly of gasoline.

The primary technique for dealing with a fuel-flooded engine is to crank with the choke off and throttle wide open. This will pump the surplus fuel through the exhaust. A handful of dry spark plugs speeds the process, as does application of compressed air through the spark-plug port.

In extreme cases, it is probably wiser to remove the spark plug and allow the engine to sit for a few hours until the surplus fuel evaporates. Even oil flooding will eventually cure itself.

Engine runs

Reading spark plugs is not difficult if you use a "weathered" plug with light carbon accumulations already in place. For the reading to be meaningful, the ignition system must function normally and the spark plug must be the correct type (heat range) for the engine. Run the engine for a few minutes at the selected speed and chop the throttle abruptly to kill the engine. Most of these tests are made at three-quarters throttle and under load.

The richer the mixture, the cooler the spark plug runs and the darker the carbon deposits. Today's small engines are set up to run slightly rich to keep cylinder temperatures under control. Spark-plug tips appear darker than they would in an automobile with emission controls.

The correct mixture will look like the color of coffee with a dash of cream (dark brown) on the spark-plug tip. When you make carburetor adjustments (see chapter 4), err on the side of richness, especially at idle and under high-speed, no-load conditions.

A lean mixture will bleach the spark-plug tip white. Other symptoms include a pop back through the carburetor under sudden acceleration or load and a lean roll as the engine drops rpm and loses power. A lean mixture might be caused by an improperly adjusted carburetor (when provision for adjustment has been provided), a partially clogged carburetor and/or fuel filter, or an air leak upstream of the throttle plate.

With a rich mixture, the spark-plug tip will be coated with black deposits. In extreme cases, the carbon deposits might have a loose, fluffy texture and smell of unburned fuel. Secondary symptoms include an acrid exhaust smell, possible black smoke in exhaust, and a "soft" exhaust note. Carbon deposits might collect at muffler discharge. Excessive fuel delivery can be traced to a dirty air filter, partially engaged choke, improperly adjusted carburetor, or an internal malfunction in float-type carburetors. Again, please note that an ignition misfire produces the same symptoms as an overly rich mixture.

Driven equipment

Sometimes it is easy to forget that the engine and the equipment is coupled to constitute a system. Worn equipment bearings, overly tight vee-belts, or severe shaft misalignments translate into difficult starting, loss of performance, excessive fuel consumption, and possible engine overheating. A loose drive coupling, belt, or rotary mower blade can make itself known by kickback, such as when the engine tries to pull the starter cord from your hand.

3

Ignition systems

Older Briggs & Stratton single-cylinder engines were fired by magnetos of four distinct types, each designed in different eras and with few interchangeable parts. Newer engines use variants of the Magnetron breakerless system. These solid-state units are manufactured in two basic styles and in kit form for retrofitting most magneto-equipped engines.

System troubleshooting

In terms of frequency of failure, ignition system components rank as follows:

1. spark plug
2. breaker points (magneto)
3. flywheel key (magneto and Magnetron)
4. condenser (magneto)
5. ignition coil (magneto)
6. Magnetron
7. under-flywheel and external wiring

Spark plugs

Spark plugs are sacrificial items replaced at the first sign of hard starting or misfiring. Recommended spark plugs are shown in Table 3-1. Engines leave the factory with Champion resistor-type spark plugs installed. The J19LM and its resistor twin are "hot" plugs, standard for two-cycle engines and sometimes fitted to side-valve engines with a history of spark-plug fouling. This condition can occur in well-tuned engines that operate for long periods under light loads. Although I do not have scientific proof, it appears that full-sized plugs (¹³⁄₁₆-in. wrench flats) last longer than the ¾-in. RCJ8 "peanut" variety.

Table 3-1.
Spark-plug recommendations

Length	Type
1.5 in.	2.0 in.
CJ-8	
RCJ-8	RJ-8 Champion (resistor)
235	295 Autolite
245	306 Autolite (resistor)
Two-Cycle Std., Side-Valve Optional	
J19LM	Champion
RJ19LM	Champion (resistor)
OHV Models	
RC12YC	Champion (resistor, long reach)
3924	Autolite (resistor, long reach)

All spark plugs are gapped at 0.030 in., a chore that one of the wedge-type gauges shown in Fig. 3-1 makes easier. Insert the gauge between the center (fixed) electrode and the side (adjustable) electrode and rotate the gauge to the desired setting.

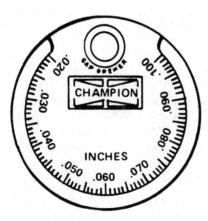

3-1 *Champion spark-plug gauge, available at auto-parts stores.*

The interface between the spark-plug gasket and the cylinder head acts as a sink to conduct heat away from the spark-plug tip. Remove all traces of oil and grease from this area. Some mechanics apply a squirt of silicone—a moderately effective lubricant—to the spark-plug threads prior to assembly. Run in the new plug three turns by hand and torque to specification (Table 3-2).

Table 3-2.
Flywheel and ignition system specifications*

Model	Armature air gap** (inch)	Spark-plug torque (lb/ft)	Flywheel nut torque (lb/ft)
60000, 80000, 90000, 100700, 110000, 120000	0.006–0.010	18–22	55
Two-cycle			
95700, 96700	0.009–0.016	13.5	30
Europa			
97700, 99700	0.006–0.014	18–22	60
104700 (ohv)	0.008–0.012	13–18	65
100200, 100900, 130000, 170000, 190000, 220000, 250000	0.010–0.014	18–22	65
Cast-iron block			
230000, 240000	0.010–0.014	18–22	145

  * Single-cylinder engines, except Vanguard and two-cycle Sno-Gard.
 ** Two-legged armatures only, magneto and Magnetron®. Obsolete three-legged magneto armatures were gapped at 0.014 in. for 6- and 8-CID engines; 0.014 in. for 10- and 13-CID engines; and 0.024 in. for iron-block 23-CID models.

Flywheels

With the exception of the rarely encountered Magna-Matic magneto, conventional and solid-state Briggs & Stratton ignition systems receive firing pulses from magnets cast into the rim of the flywheel. The geometric relationship between the coil armature and flywheel magnets affects engine timing and spark intensity.

To access the flywheel, it is necessary to remove the cooling shroud. Depending on the engine model, this might entail removing the rewind starter, fuel tank, carburetor, throttle cable, and miscellaneous hardware. The flywheel is secured to the crankshaft by a nut, lock washer, aluminum key, and taper fit.

Warning: Disconnect and secure the spark-plug lead before servicing the flywheel. Also disconnect the positive battery cable when an electric starter is fitted.

The starter clutch doubles as the flywheel nut on engines that use the traditional Briggs rewind starter (recognized by the square drive shaft). The clutch requires a special wrench PN 19114 or, preferably, PN 19161 (Fig. 3-2). The latter can be used with a torque wrench during assembly. Engines equipped with modern rewind starters and iron-block engines that still use a notched pulley for starting, have their flywheels secured with a hex nut.

Note: Some engines not equipped with rewind starters have left-hand crankshaft threads. Determine the lay of the threads (which will be visible) before attempting to undo the nut.

It is also necessary to prevent the flywheel from turning when the nut is loosened or tightened. Lawnmower mechanics wedge the blade against the mower deck with a short length of a 2×4. A wooden block, jammed against a flywheel fin and the bench, works for cast-iron flywheels. If the threads are rusted, tie the engine down with C-clamps.

Aluminum wheels are too fragile to secure by a single fin. Briggs makes a special tool for 6.75 in. and smaller flywheels, which spreads the force over two fins. A strap wrench, available from large hardware stores and from the factory as PN 19372 fits all wheels (Fig. 3-3). Conventional split-type and Bellville-dished lock washers are used. Note that Bellville washers install with the concave side next to the flywheel.

Caution: Do not attempt to hold a flywheel by inserting a screwdriver between the impeller blades.

Heavy, cast-iron flywheels can be lifted without special tools. Insert two large screwdrivers between the inboard side of the flywheel and the engine block. Holding the tools in position with your right hand, "stiff arm" the screwdrivers with the open palm of your left hand. This method will usually suffice. In stubborn cases, shock the wheel with directed hammer blows on the rim. Avoid striking the magnets.

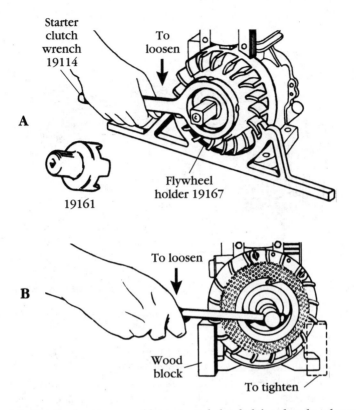

3-2 *Flywheels are secured by a started clutch (A) or hex-head nut. (B) If you have to buy the tool, torque-wrench adaptable socket PN 19161 is a better choice than PN 19114. Note that cast-iron block engines, fitted with notched pulleys in lieu of rewind starters, use left-hand threads at the flywheel.* Briggs & Stratton Corp.

Aluminum wheels should be removed with a special puller that threads into holes cast into the hub. A conventional gear puller of the kind that grips the flywheel rim will distort, and might break, the fragile casting. The factory supplies hub pullers for all sizes of aluminum wheels:

Puller number	Engine model range
PN 19069	60000 through 120000
PN 19165	140000, 170000, 190000, 1975, and earlier 250000
PN 19203	220000, 230000, 240000, post-1975 250000 through 320000

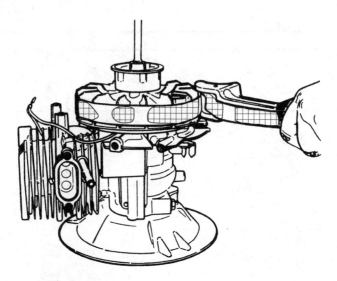

3-3 *A strap wrench, used here on a Tecumseh engine, has almost universal application.*

There is a "bootleg" way of freeing the flywheel from its taper, mentioned because it is almost universal in the trade. A steel bar threaded over the crankshaft and struck with a hammer will impart enough energy to shock the wheel loose. Side effects include possible crankshaft breakage, demagnetization, and damage to ball or roller main bearings. Briggs & Stratton could, in theory, void the warranty on an engine that has been treated in this fashion.

Other manufacturers, perhaps bowing to the inevitable, accept the use of knockers and even supply them. Tecumseh PN 670103 fits ½-in. 20-TPI (threads per inch) shafts; PN 670118 is the left-hand variant of the same thread; PN 670169 fits ⁷⁄₁₆-in. 20-TPI shafts.

Warning: Wear eye protection when using a knocker or, indeed, whenever hammering steel. Hit squarely and as hard as necessary to shock the wheel off. One serious blow probably does less damage to the magnets than a series of ineffectual taps.

Once the flywheel is lifted, carefully examine the hub area for cracks that usually begin at keyway corners (Fig. 3-4).

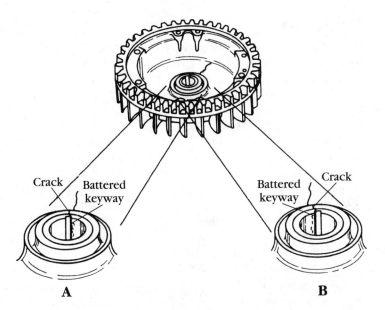

3-4 *A crack or wallow on the leading edge of the keyway (closest to the direction of rotation) suggests that the flywheel was loose and overrun by the crankshaft. (A) This condition can usually be ascribed to a loose flywheel nut. Damage to the trailing edge of the keyway means the flywheel attempted to overtake the crankshaft. (B) That could only have occurred if the crankshaft were suddenly stopped, as when a rotary lawnmower blade strikes an obstacle. In either case, the flywheel (and possibly, the crankshaft) must be replaced.*

Warning: Serious injury could result from operating an engine with a cracked or otherwise damaged flywheel.

Using a new key as a gauge, inspect the flywheel and crankshaft keyways. Some "slop" is permissible, but wallow, serious wear, or impact damage means that the affected part must be replaced. Replace the original key if dented or visibly worn.

Caution: Use the appropriate key, PN 222698, for most engines; PN 800036 for the two-cycle Model 96700. Substituting a steel key for the aluminum factory part could result in expensive damage to the flywheel and/or crankshaft.

Magnetos

Briggs & Stratton used four magnetos, three of which are discussed in this section. I have never seen the fourth magneto, a Magnavac, which has magnetically tripped points encased with the coil assembly.

Theory of operation

Servicing magnetos requires some general understanding of how they work, if only to improve the odds of purchasing the correct replacement part. Magnetos convert the energy of spinning flywheel magnets into voltage. Two basic principles are involved:

- Voltage is impressed upon a conductor when the conductor is subjected to a moving magnetic field.
- Current flow through a conductor induces a magnetic field at right angles to the conductor.

Magnetos consist of two circuits, known as the primary and the secondary. Part of the primary circuit is visible as the insulated wire that runs from the "hot" (or insulated) breaker point to the ignition coil and the kill switch (Fig. 3-5). If you were to open the coil, you would see that the primary makes some 200 turns around the armature laminations before going to engine ground.

The secondary, consisting of 10,000 or more turns of extremely fine wire, is wound over the primary and terminates at the spark plug. Most Briggs & Stratton coils are externally grounded, as shown in Fig. 3-5.

Figures 3-6A through C illustrate magneto operation about as well as pictures can. In Fig. 3-6A, the rim magnet field saturates the coil, inducing voltage that goes to ground through the closed contact points. In Fig. 3-6B, the flywheel has rotated a few degrees causing the magnetic activity and primary current to peak. What has occurred to this point reflects the first of the two principles covered earlier: voltage is impressed upon a conductor when the conductor is subjected to a moving magnetic field.

Figure 3-6C depicts activity a few crankshaft degrees later, at the moment of point break. Once the points separate, the primary circuit loses ground and current flow abruptly ceases.

The second principle states that current flow through a conductor induces a magnetic field at right angles to the conductor. When the points open, the magnetic field associated with the

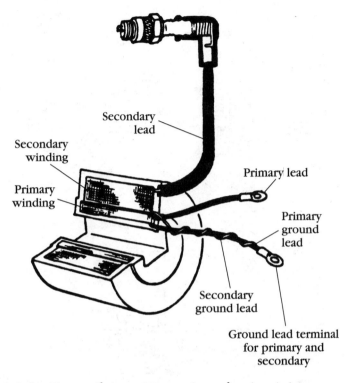

3-5 *Ignition coil in cutaway view, showing primary and secondary windings. The ground lead is not present on internally grounded coils.* Tecumseh Products Co.

primary current collapses on itself to induce a high voltage (proportional to the turns ratio) in the secondary. This voltage arcs to ground at the spark plug.

Oversimplifying somewhat, the condenser (or capacitor) acts as a buffer and stores electrons that would otherwise arc across the points as they separate. Burnt points are the classic symptom of condenser failure.

The kill switch (shown in Fig. 3-6A) is a normally open switch that grounds the primary when tripped. Key-operated ignition switches ground in the *off* position.

Point set & condensers

Service operations center around the point set and condenser, which should be replaced as an assembly during tune-ups and

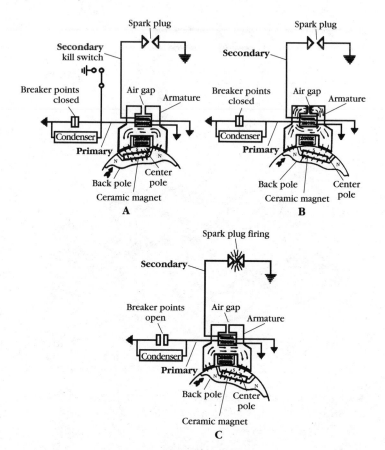

3-6 *Magnets imbedded in the flywheel rim move into proximity with the ignition coil, saturating the primary windings with magnetic energy. Current flows in the primary circuit, returning to the coil through the closed (and grounded) ignition points. (A) Primary current creates its own magnetic field that saturates the secondary windings. (B) At peak primary-current flow and maximum secondary saturation, the contact points open. (C) Collapse of the magnetic field induces a high-voltage surge in the secondary circuit that, seeking ground, arcs across the spark-plug electrodes.*

whenever you encounter a magneto fault that a new spark plug and flywheel key does not cure.

Normal modes of failure include oxidation (gray, highly resistive deposits on the tungsten contacts), overheating (blackened contacts, with possible discoloration of the moveable arm), metal migration (one contact pits, the other gains metal), and loss of gap. Except for the last, which results from wear on the plunger or a slipped adjustment, all of these faults implicate the condenser.

Before you begin, note the routing of wires from the point set to the ignition coil and to the wiring-harness connection point. The routing should be duplicated exactly upon assembly to preclude damage to the insulation from contact with the flywheel and other moving parts.

Remove the old contacts and condenser and note the arrangement of parts. The assembly drawings in this chapter are useful as general guides, but do not take into account running production changes. Remove all traces of oil from the point set and condenser mounting surfaces.

Briggs & Stratton magnetos employ a 0.020-in. point gap. Adjust as follows:

1. Turn the crankshaft until the point plunger retracts and contact points touch. It might be necessary to make a preliminary adjustment to bring the points together.
2. Turn the crankshaft until the points open to their widest extension.
3. Determine the distance between the moveable and fixed contact with a 0.020-in. flat-bladed feeler gauge. (Some mechanics bracket the adjustment with 0.021- and 0.019-in. blades.)
4. Move the stationary contact as necessary to establish the gap. The mechanism for moving the contact varies with the magneto type.

Normally, magneto service ends when the spark plug, flywheel key, point set, and condenser have been replaced. Further repairs are driven by spark testing.

With the spark plug removed and PN 19161 connected between the spark-plug lead and engine ground, install the flywheel key and slip the flywheel over the crankshaft. It is not necessary to secure the wheel at this stage. Spin the flywheel vigorously by hand, exerting care to avoid contact with the governor vane and other sharp-edged parts. The magneto should deliver a thick, blue spark.

Failure to deliver a quality spark means:

• point contact faces are contaminated with oil;
• point contact faces are oxidized;

- the point set or condenser is faulty; and
- another part of the system is failed.

Wiping the feeler gauge with your fingers can transfer enough oil to prevent ignition. Spray the contacts with one of the aerosol cleaners sold for this purpose and test for spark.

Oxidation might not be visible to the eye. Burnish the points with a business card and retest.

If the spark remains nonexistent or marginal, replace the newly installed point set and condenser with known good components. Next, trace the wiring. Look for loose connections and frayed insulation as described in the following section: "External circuits." As a last resort, replace the coil.

Under-flywheel magnetos

The under-flywheel magneto is the classic Briggs & Stratton magneto, recognized by its under-flywheel, plunger-operated point set. Millions of these units remain in service on engines with displacements of up to 17 cubic inches.

Three versions were built, distinguished by the point set. The most popular, shown in Fig. 3-7, combines the stationary contact point with the condenser. This arrangement reverses the usual order of things by making the moveable contact the ground. The second version, used on many horizontal shaft engines, has an entirely conventional point set with the condenser wired to the moveable contact. Few mechanics still working have seen the third type, which differs in minor respects from the second and does not have the dust cover common to the others. All versions trigger off a flat milled on the crankshaft to generate a spark with every revolution.

Figure 3-8 illustrates the two most current point sets. Figure 3-9 describes the connector used to secure primary wire to the condenser/stationary contact. Note the depressor tool used to compress the spring. These tools come packaged with factory-supplied replacement point sets. In a pinch, a pair of long-nosed pliers can be used to rotate the spring "threading" the wire on and off.

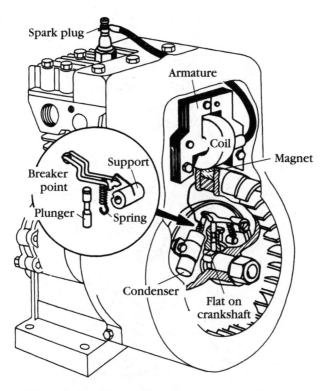

Spark plug
Armature
Coil
Magnet
Support
Breaker point
Plunger
Spring
Condenser
Flat on crankshaft

3-7 *Millions of Briggs & Stratton under-flywheel magnetos continue to give service. The most popular variant employs the point-and-condenser set shown.*

It is not unusual for the point set to be covered with a light film of oil, especially on vertical shaft engines, but puddles of oil mean upper main-bearing seal failure. This condition must be corrected if the magneto is to work reliably.

A worn plunger and/or plunger bore can also contribute to point oiling, but the fault is rarely encountered and probably not worth repairing on engines that accept a Magnetron retrofit. The plunger reject length is 0.870 in.

Caution: Install the plunger with the grooved end toward the breaker points. Reversed installation will result in point oiling.

The adjustment procedure is pictured in Figs. 3-10 and 3-11.

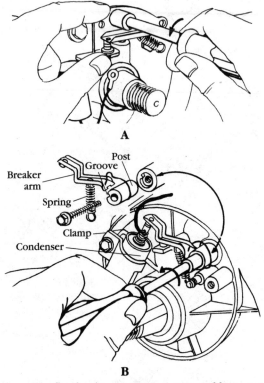

3-8 *Both under-flywheel point sets are secured by a single screw. The older two-piece set employs a remote condenser, wired to the moveable arm. (A) The newer point set (B) includes a braided ground wire, which should be assembled over the top of the post, as shown.*

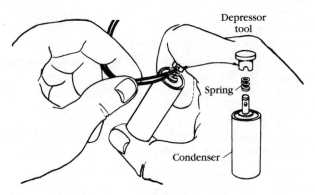

3-9 *A depressor tool, packed with Briggs & Stratton replacement point sets, should be in every mechanic's toolbox.*

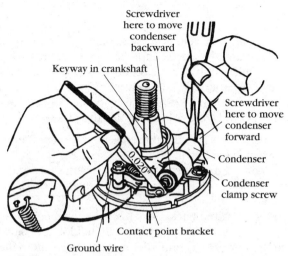

3-10 *Assemble the integral point-and-condenser set as shown, with the braided ground wire outboard of the point-arm post and the open eyelet of the point spring attached to the moveable point arm. Turn the crankshaft to bring the keyway adjacent to the point plunger, snug down the condenser-clamp screw and set the gap to 0.020 in. Use a screwdriver to move the condenser as necessary. Tighten the condenser-clamp screw and recheck the gap, which almost invariably, will have changed. Repeat the process, this time compensating for the effect of the clamp screw.*
Briggs & Stratton Corp.

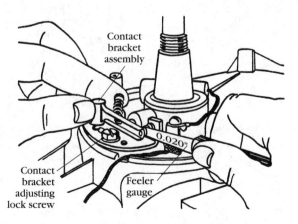

3-11 *Briggs' two-piece point sets have a screwdriver slot to facilitate adjustment. Otherwise, the drill is as described in the caption for Fig. 3-10.*

Coils

Late production coils employ U-shaped, "two-legged" armatures, as illustrated by the example in Fig. 3-12. "Three-legged"
armatures, which look like the letter "E," go back many years,
but are still occasionally encountered. Table 3-2 lists air-gap
specifications for both types. The air gap must be set whenever
you install a coil. Figure 3-12 illustrates the process.

1. Obtain shim stock of the prerequisite thickness (an old
   feeler gauge blade will do).
2. Rotate the magnets away from the armature legs and
   loosen, but do not remove, the two armature holddown
   capscrews.
3. Push the armature back away from the flywheel and
   tighten the capscrews to temporarily hold it in position.
4. Turn the flywheel to bring the magnets adjacent to the leg
   ends, or poles.
5. Insert the shim and loosen the holddown screws. Attracted
   by the magnets, the armature will snap against the
   flywheel.
6. Tighten the capscrews and remove the shim. The gap is
   now set.

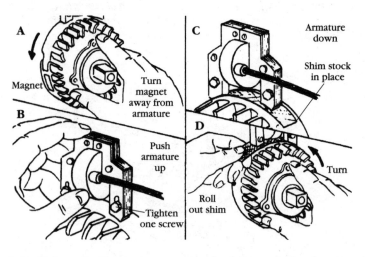

3-12 *Briggs & Stratton recommends this method of setting
magneto and Magnetron armature air gaps. A nonmagnetic
feeler gauge of the type sold for automotive air-conditioning
work gives equivalent results.*

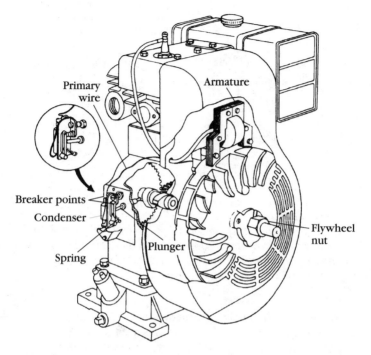

3-13 *The external point set was used on horizontal shaft engines in the 19 to 32 cubic in. displacement range.* Briggs & Stratton Corp.

Note: The air-gap specification is an approximate number that includes a cushion for manufacturing inaccuracies, main-bearing wear, and crankshaft eccentricity. Adjust to the narrow end of the specification range, allowing a few thousandths of an inch for high speed runout. In no case can the flywheel be allowed to rub the armature.

External point-set magneto

External point-set magnetos are most often seen on models 193000, 200000, 233000, 243000, 300000, and 320000, and back to the days of 19D and 23D engines, these magnetos trigger from the camshaft (Fig. 3-13).

Point-set replacement Using Fig. 3-14 as a reference, follow this procedure to replace the point set:

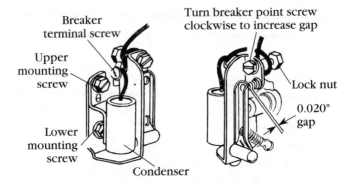

3-14 *Oil intrusion has been a problem with these magnetos. Point assembly holddown screws should be coated with a non-hardening sealant and the plunger oil seal (suggested in the drawing as the cylindrical object below the locknut) should be functional. An oil seal can easily be retrofitted to early production engines that do not have this feature.* Briggs & Stratton Corp.

1. Remove the point compartment cover.
2. Turn the flywheel until the contacts are fully open.
3. Remove the upper mounting screw and lift out the condenser.
4. Remove the lower mounting screw, now accessible.
5. Slack off the locknut and back out the breaker adjustment screw to release the point assembly.
6. Disconnect the coil lead at the breaker terminal screw.

To assemble:

1. Coat mounting screw and adjusting-screw threads with a nonhardening sealant to prevent oil entry into the point compartment.
2. Position the point set and lightly run in the lower mounting screw that passes through the plunger oil-seal retainer.
3. Align the upper hole in the oil-seal retainer with its threaded boss and start the upper mounting screw.
4. Tighten the lower mounting screw.
5. With the points fully open, turn the adjustment screw as necessary to achieve a 0.020-in. gap between contact faces. Snug down the locknut and recheck the adjustment.
6. Install the point compartment cover. The notch cut in the cover for the primary wire can be sealed with auto-body putty or an equivalent.

Timing The timing procedure is a fairly complex process, the object of which is to index a timing mark on the coil armature with a mark on the flywheel at the moment of point break. An ohmmeter is required.

1. Verify that the point gap is exactly 0.020 in.
2. Position the flywheel over the crankshaft stub and insert the key.
3. Lightly tighten the flywheel.
4. Rotate the flywheel clockwise to bring the piston within an inch or so of top dead center (tdc) on the compression stroke.
5. Connect an ohmmeter across the moveable ("hot") point arm and engine ground. Because the points are still closed the meter will register zero resistance.
6. Inch the flywheel clockwise, while observing the ohmmeter. Stop when the meter first registers resistance. The reading will be on the order of three or four ohms, which is the resistance of the coil primary winding. Should you miss the moment of point break, continue to rotate the flywheel clockwise until the points break again.

If the arrow embossed on the flywheel aligns with the arrow on the coil bracket, the engine is in time, and you can proceed with assembly. If the arrows do not align, you must remove the flywheel—without turning the crankshaft—and reposition the coil bracket, which is secured by four cap screws, two of which are under the flywheel (Fig. 3-15).

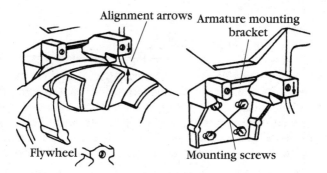

3-15 *The timing drill for the external-point magneto centers on the armature bracket, which is accessible when the flywheel is removed.* Briggs & Stratton Corp.

1. Loosen the armature bracket screws.
2. Mount the flywheel and key, again exerting care not to disturb the crankshaft. Torque the flywheel nut to specification (Table 3-2).
3. Gently tap the armature to bring the arrows into alignment.
4. Rotate the flywheel through two revolutions and recheck the timing.
5. The coil is flanked by two screws that secure the coil and armature assembly to the bracket. Loosen these screws and set the air gap as shown in Fig. 3-16.

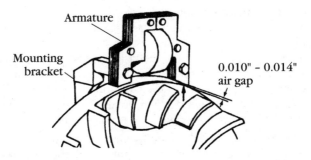

3-16 *Shim stock can be used to set the air gap as shown. Rotate the flywheel to bring the magnets under the armature, release the armature at its two holddown screws, tighten, and roll out the shim stock.* Briggs & Stratton Corp.

Magna-Matic

The Magna-Matic was used on models 7, 14, 19, 191000, 23, and 231000. By all reasonable reckoning, it is an artifact of historical interest as the most complicated, service-intensive magneto Briggs ever built (Fig. 3-17). The information that follows is for those readers who might have an otherwise sound engine with a failed Magna-Matic and no place to go for help. Most mechanics are unfamiliar with this magneto and the factory has dropped all reference to it from the service literature. In the event that parts unavailability makes repair impossible, the unit can be replaced with a Magnatron pulse generator and matching flywheel.

Points replacement   Replacement point sets (PN 291447) and condensers (PN 291364) can usually be had from your regional distributor. Become familiar with Fig. 3-18A to C before proceeding.

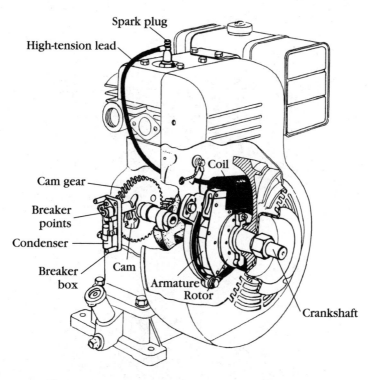

3-17 *The Magna-Matic can be recognized by the rotor and horseshoe-shaped coil armature.* Briggs & Stratton Corp.

To disassemble:

1. Remove the terminal screw.
2. Remove the spring screw and release tension on the moveable point arm.
3. Undo the breaker shaft nut and turn it counterclockwise until flush with the end of the breaker shaft.
4. Lightly tap the nut to free the point assembly from the tapered breaker shaft.
5. Remove the breaker shaft nut and lock washer.
6. Lift out the moveable point arm, together with the breaker plate.

To reassemble:

1. Position the breaker plate over the insulating plate with the detent on the underside of the breaker plate aligned with the matching hole in the insulating plate (Fig. 3-18B).

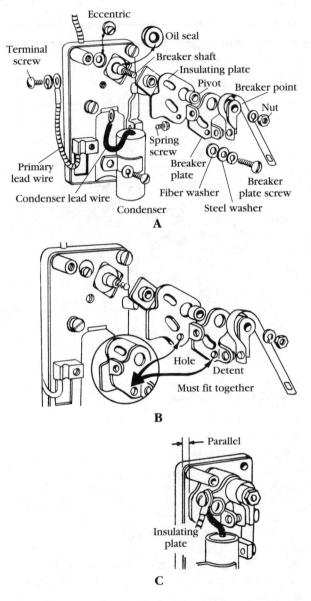

3-18 *For shear complexity, the Magna-Matic point set cannot be beat. (A) Install the breaker plate with the detent in the hole provided in the insulating plate. (B) Turn the eccentric screw to align the left-hand edge of the insulating plate with the left edge of the breaker box (C).*

Incorrect assembly will warp the breaker plate and cause
point misalignment.

2. Snug the breaker plate screw tightly, enough to hold the
   plate in position but not so tight that the plate cannot be
   moved.
3. Turn the eccentric screw to bring the left-hand edge of
   the breaker plate parallel to the left side of the breaker
   box (Fig. 3-18C).
4. Turn the breaker shaft clockwise to its limit of travel.
5. While holding the shaft against its stop, slip the moveable
   point arm over the shaft. Place the lock washer over the
   end of the shaft and tighten the shaft nut.
6. Bring the breaker points up on the cam to their fully
   open position.
7. Turn the eccentric screw to obtain a 0.020-in. gap
   between the contacts.
8. Tighten the breaker plate screw and recheck the point gap.
9. Burnish the points with a business card.
10. Check for spark.

Rotor   Install late-production rotors with 0.025-in. clearance
between the inner face of the rotor and the main bearing
housing (Fig. 3-19). The slot in the clamp should be centered
between rotor slots.

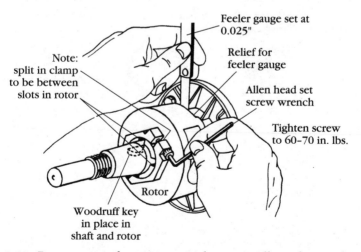

3-19 *Rotor to main bearing cover clearance allows the crankshaft to "float" without damaging the rotor. First-production
rotors were fixed with a set screw and did not require
adjustment.* Briggs & Stratton Corp.

Note: The clamp acts as a keeper to complete the magnetic circuit between rotor poles. Clamp and rotor must remain together during storage.

The rotor is inscribed with three timing marks, identified by engine displacement. The appropriate mark must align with the arrow on the armature at the moment of point break. Time as for the external point set magneto previously discussed.

Check the rotor/armature clearance at all points on the rotor periphery, using a 0.5-in.-wide, 0.004-in. feeler gauge. If the gauge binds at any check point, either the main bearings are worn or the crankshaft is bent. In either case, major repairs are in order.

Magnetron

The Magnetron pulse generator went into limited production in August 1982, and has since become the new standard, replacing all magnetos in the line (Fig. 3-20). The device consists of a conventional ignition coil and a transistorized switch, triggered by the flywheel magnets. Early models employed a replaceable switch module. Recently designed models integrate the switching circuitry and coil into a single assembly.

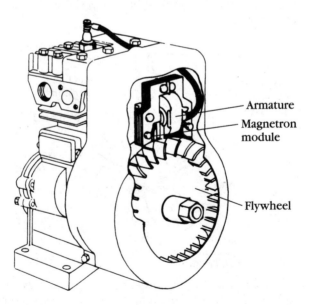

3-20 *Magnetron has now become standard on Briggs & Stratton engines.*

Troubleshooting

Two-piece Magnetrons offer a point of attack for the technician who can substitute another switch module for the original. Odds are it is the module, and not the coil, that is at fault. Magnetrons with nonreplaceable switch modules are essentially throwaway items.

Although apparatus exists to test these systems, most mechanics prefer to test by substitution, if necessary borrowing one from another engine. Table 3-3 identifies Magnetron systems most often encountered on single-cylinder, non-Vanguard engines.

Table 3-3.
Magnetron applications

Application	Magnetron Armature PN
2 thru 4-hp horizontal & vertical	398593
5-hp horizontal & vertical	397358
4-hp two-cycle, 5-hp Quantum, & 5-hp Europa	802574
7 thru 16-hp horizontal & vertical	398811

Switch module

One switch module fits all two-piece Magnetron units and can be used to update most magneto systems to the newer technology.

Magneto to Magnetron conversion The switch module kit replaces breaker points and condenser in nearly all aluminum-block engines with "two-legged" armatures.

PN 394970, consisting of the module and miscellaneous parts, costs about $20 and usually takes a mechanic about an hour to install the first time around.

1. Remove the spark plug and disconnect the positive battery cable at the battery on electric-start models.
2. Lift the flywheel to access the primary and kill-switch wiring. You may leave the points and condenser in place and snip the wires at the dust cover (Fig. 3-21). If you

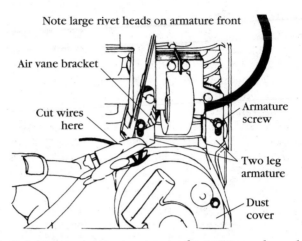

Note large rivet heads on armature front

Air vane bracket

Cut wires here

Armature screw

Two leg armature

Dust cover

3-21 *It is not necessary to remove the points and condenser when retrofitting a Magnetron module. Note that large rivet heads define the outboard, or flywheel, side of the armature.*
Briggs & Stratton Corp.

want to make a neat job of it, remove the points and condenser and seal the point plunger hole with PN 231143. Failure to plug the hole will result in a major oil leak.

3. Remove the two cap screws that hold the ignition coil armature and, in many applications, the governor vane bracket to the engine. You might have to cut part of the bracket away for module clearance as shown by the shaded area in Fig. 3-21.

4. Stand the coil armature upright on its "legs" with the engine side facing you. The heads of the rivets that hold the laminations together should not be visible. As the armature is now oriented, the Magnetron module installs between the coil and the right leg.

5. Slip the module into position without forcing it. Clip the plastic hook over the armature shoulder (Fig. 3-22).

6. Prepare to make the electrical hookup. One or more hot (insulated) wires originally ran from the point assembly to the coil and to external engine shutdowns. The coil was grounded with a bare wire connected to one of the armature holddown screws. Two wires come off the module: a long ground wire with a terminal on its end and a shorter hot wire.

7. Peel back the insulation ¾ inch from the ends of the wires that originally ran to the points. Scrape the varnish from the wire ends.

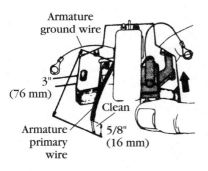

Armature
ground wire

3"
(76 mm)

Clean

Armature
primary
wire

5/8"
(16 mm)

3-22 *The module installs
on the right-hand leg, as
viewed from the under-
and rivetless-side of the
armature.* Briggs & Stratton Corp.

8. Route the wires between the ignition coil and the
cylinder. Wires should be long enough to reach the
module. If not, make a splice with primary wire supplied
in the kit. Use only as much wire as necessary to make
the connection and allow room for installation. Twist the
wires together two turns, and solder using 60/40 rosin-core
solder. Insulate with heat-shrink tubing. **Caution:** Do not
use acid-core solder or crimp-on connectors.

9. Solder these wires to the short Magnetron hot wire as
described in the previous step. Work quickly. Use a
minimum amount of solder to avoid heat damage to the
module.

10. The kit includes a small coil spring and a T-shaped
retainer with a hook on the end. Slip the spring over the
long arm of the T, so that it shoulders on the crossbar
(Fig. 3-23). Place the retainer, hook first, into the recess
on the side of the module.

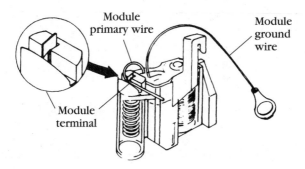

Module
primary wire

Module
ground
wire

Module
terminal

3-23 *A clip-and-spring assembly secures wires to the module.*
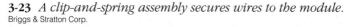
Briggs & Stratton Corp.

11. Compress the retainer spring with a Briggs & Stratton condenser or drill bit, as shown in Fig. 3-24. Insert the wire bundle, which you soldered together in Step 9, under the head of the retainer. Release the spring and, using long-nosed pliers, turn the retainer as necessary to engage the hook into the slot. Trim wire ends about ³⁄₁₆ inch out from the retainer.

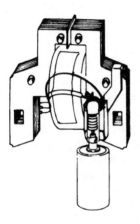

3-24 *Twist and solder primary (in-sulated) wires to the module "hot" (noninsulated) wire. Open the retainer with a punch, drill bit, or Briggs & Stratton PN 294628 condenser.*

12. Twist the module-ground and armature-ground wires together, near the armature. Solder the connection. Since both wires have connectors, the shorter of the two may be snipped off.
13. To prevent vibration damage, plaster the wires to the underside of the coil with Permatex No. 2 (Fig. 3-25).
14. Install the armature and, when present, the air-vane bracket on the engine. Use the holddown screw on the right side of the coil armature as the ground connection. Do not attach the armature/module ground to the air-vane bracket.
15. Set the armature air gap, as described under the preceding "Coil" heading and to the specification in Table 3-2.
16. Use the key provided in the kit to slip the flywheel over the crankshaft stub. Connect ignition tester PN 19051 between the spark-plug cable terminal and ground. Spin the flywheel vigorously by hand to attain the 350 rpm necessary for Magnetron spark production.
    Warning: Contact with the air-vane bracket can result in a nasty cut.

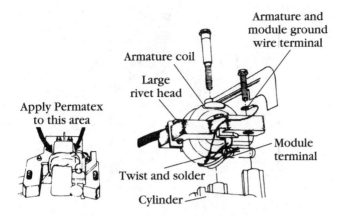

3-25 *Secure the wires to the underside of the coil with sealant. Note that the module and coil ground wire attach to the armature holddown screw on the leg opposite the air-vane bracket.*
Briggs & Stratton Corp.

17. If there is no spark, verify that the wires are connected properly—primary to module hot wire and armature ground to module ground—and check for cold (incompletely fused) solder joints.
18. Complete the assembly, torque the flywheel to specification, and test the engine.

Replacing an existing switch module   The replacement procedure follows the previous outline except that breaker points and condenser are, of course, not part of the picture. Be careful not to overheat the module retainer when unsoldering the wires (Fig. 3-26).

Shutdown switch & interlocks

Modern practice is to provide single-lever control of engine speed, choke engagement, and ignition. Briggs & Stratton shutdown switches consist of a copper tang, insulated from engine ground and connected to the primary side of the ignition system. The throttle lever is grounded. Moving the lever past the idle detent brings it into contact with the tang, shorting the ignition (Fig. 3-27).

Failure to stop the engine is usually caused by an improperly adjusted control cable that holds the lever away from the tang. Models with lever-engaged chokes usually require some judi-

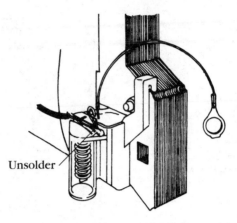

3-26 *Exert extreme care when unsoldering wires from a suspect module.* Briggs & Stratton Corp.

cious bending of the choke link to enable the choke to come full on at one extreme of lever travel and the switch to function at the other.

Also check for a broken or detached primary wire, wear on the switch tang, and a loose pivot on the throttle cable that might introduce lost motion into the system. The PN 297472 stop switch has been standardized across most model lines.

Ignition interlocks, or safety switches, are included in the primary circuit to protect operators. The most common of these is the "deadman's switch" that shorts the ignition on rotary lawnmower engines when the operator releases the handlebar. At the same time, a spring-loaded flywheel brake engages and quickly brings the engine to a halt. Some Briggs & Stratton engines employ a second PN 296472 stop switch, which is located on the brake assembly and tripped by the same mechanism that releases the brake. Others make use of the existing stop switch, which trips when the throttle lever is moved past idle and when the operator lifts his hands from the handlebar. Figure 3-27 illustrates the test procedure for the latter configuration.

Interlocks can become quite complex when the equipment manufacturer has a hand in their design. Some include logic modules to defeat tampering. From a mechanic's point of view, interlocks fall into two categories: those that are normally open (NO) and those that are normally closed (NC). NO interlocks shunt primary voltage to ground when activated. NC interlocks open to interrupt the primary circuit. In either case, the engine stops.

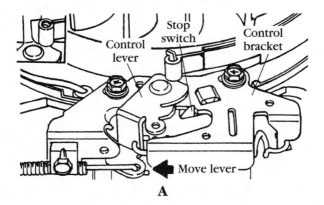

A

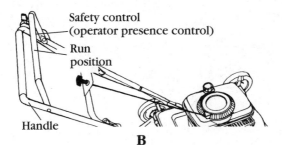

B

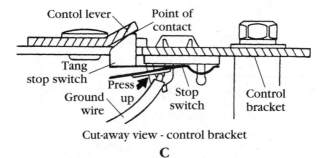

Cut-away view - control bracket

C

3-27 *One approach Briggs & Stratton took to the "compliance engine" was to integrate the handlebar lever with the existing stop switch and throttle-control lever. To test, place the throttle-control lever in the "run" position (A), release the handlebar (B), and verify that the control lever moves into contact with the stop switch.*

The surest way to test interlocks is with an ohmmeter. Isolate the suspect component from the circuit and measure its resistance in the normal and tripped modes. Figure 3-28 illustrates another technique, used for simple dc circuits without antitampering devices. If an engine refuses to start with interlocks in the circuit and runs with a jumper connected as shown, the source of the problem is obvious.

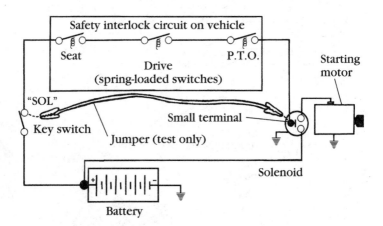

3-28 *Typical safety interlocks.*

4

The fuel system

The fuel system consists of the carburetor, air cleaner assembly, fuel line, tank, and miscellaneous fittings. Optional elements include a fuel pump (which might be integral with the carburetor), shutoff valve, and filter.

How carburetors work

We think of an engine as a source of power. From the fuel system's point of view, the engine is a vacuum pump. The partial vacuum created by the piston during the intake stroke sets up a pressure differential across the carburetor. Air and fuel, impelled by atmospheric pressure, move through the instrument to equalize pressures.

Venturi & high-speed circuit

If you look through a carburetor, you'll see that the bore has an hourglass shape, with the necked-down portion located just upstream of the throttle plate. This area is known as the venturi. As much air leaves the carburetor as enters. Consequently, air velocity through the venturi must be greater than through the straight sections of the bore on either side of it. The increase in velocity is purchased at the expense of pressure.

Fuel, under atmospheric pressure, moves from the carburetor reservoir through the main jet and into the nozzle, which opens to the low-pressure, high-velocity zone created by the venturi. The jet can be fixed, as shown in Fig. 4-1A, or adjustable. In either case, the size of the jet orifice determines the strength of the mixture by regulating how much fuel passes into the venturi. The main air jet, more commonly known as the main air

59

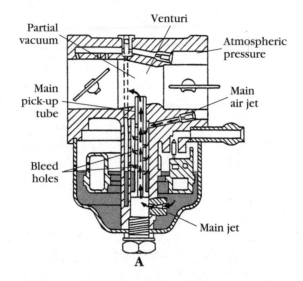

4-1 *At wide throttle angles, fuel enters the venturi through the main pick-up tube, better known as the* high-speed nozzle. *(A) An air bleed ("main air jet") emulsifies the fuel, breaking it into droplets prior to discharge.* Briggs & Stratton Corp.

bleed, emulsifies the fuel before discharge, primarily to prevent siphoning. Collectively, these parts make up the high-speed circuit—high speed because the venturi works only so long as the throttle is open. Closing the throttle blocks air flow through the venturi and shuts down the fuel circuit.

Throttle & low-speed circuit

The throttle blade, or butterfly, controls engine speed by regulating the amount of fuel and air leaving the carburetor. It functions like a gate valve, opening for the engine to develop full power and almost completely blocking the bore at idle. The restriction generates a low-pressure zone downstream of the throttle blade, exactly as if it were a venturi. Fuel enters through a series of ports drilled in the carburetor bore. The port nearest the engine—known as the primary idle port—functions when the throttle blade is against its stop (Figs. 4-1B). As the throttle cracks open, one or more secondary ports are uncovered to ease the transition between idle and main venturi startup (Figs. 4-1C). The low-speed circuit also includes an air bleed.

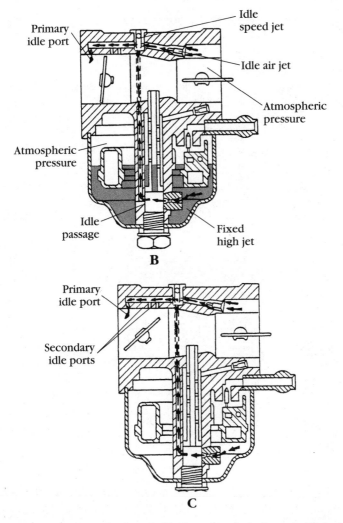

4-1 (Cont.) *When the throttle blade is closed, the low-speed circuit discharges into the primary idle port. As the blade opens, it uncovers secondary idle ports (C). Wider throttle angles generate flow through the main jet.* Briggs & Stratton Corp.

Fuel inlet

Carburetors always include a mechanism for regulating the internal fuel level, independent of delivery pressure. Most Briggs type carburetors employ a float-actuated inlet valve, known as

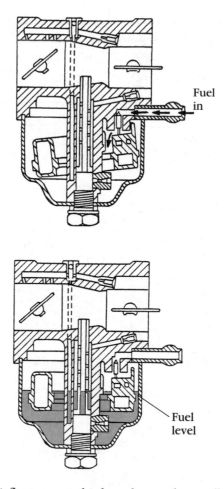

Fuel
in

Fuel
level

4-1 (Cont.) *A float-operated inlet valve, or the needle and seat, maintains a constant fuel level in the instrument (D).* Briggs & Stratton Corp.

a *needle and seat* (Fig. 4-1D). When the reservoir is full, the float forces the needle against its seat, which cuts off fuel delivery. As gasoline is consumed, the float drops, which releases the needle and opens the valve.

Suction-lift carburetors draw from the tank through a pickup tube in a manner analogous to a flit gun (Fig. 4-2). A check valve prevents fuel from draining out of the tube during starting.

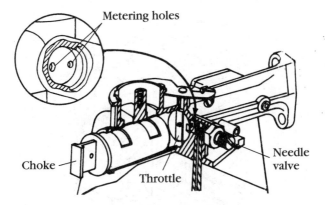

4-2 *Vacu-Jets and their Pulsa-Jet cousins draw through a pickup tube that extends into the fuel tank. An internal check ball prevents fuel from running back out of the tube. Note the plug-type choke and single-adjustment needle that controls both low- and high-speed mixture strength.* Briggs & Stratton Corp.

Cold start

With one or two exceptions, Briggs carburetors employ conventional choke valves to enrich the mixture during cold starts. The Walbro carburetor shown in Fig. 4-1A uses a pivoted choke disk. The Vacu-Jet in Fig. 4-1B has an old-fashioned plug choke. In either case, closing the choke seals off the carburetor bore. The engine, in effect, pulls on a blind pipe. All jets flow in response to the low pressure.

Figure 4-3 illustrates the Briggs automatic choke found on Vacu- and Pulsa-Jet carburetors. A spring-loaded diaphragm holds the choke closed during cranking. Upon starting, a manifold vacuum, acting on the underside of the diaphragm, overrides the spring to open the choke. The choke also operates as an enrichment valve should the engine falter under load; the loss of the manifold vacuum allows the choke to close.

Some automatic choke mechanisms are fitted with a bi-metallic helper spring. A tube connected to the breather assembly conducts warm air over the spring, which causes it to uncoil and open the choke independently of the vacuum signal acting on the choke diaphragm.

Pulsa-Prime nylon-bodied carburetors combine a fuel pickup tube with a primer pump. No choke is used. Pressing the

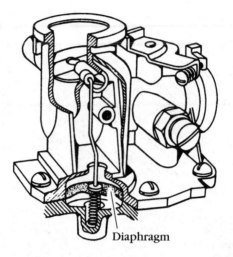

Diaphragm

4-3 *Some Vacu- and Pulsa-Jet carburetors are fitted with a vacuum-operated automatic choke. The choke butterfly should close when the engine is not running, flutter shut under sudden load and acceleration, and open at steady speed. Major causes of failure are a bent air cleaner stud, leaking diaphragm, and dirt in the butterfly pivots. The diaphragm chamber must be airtight.* Briggs & Stratton Corp.

primer bulb evacuates air from the pickup tube, which causes the level of fuel in it to rise.

External adjustments

Figure 4-4A illustrates a Flo-Jet carburetor with adjustable main and low-speed jets. Backing out the needle-tipped adjustment screws opens the jet orifices and enriches the mixture (Fig. 4-4B). Tightening the screws makes the mixture leaner by restricting the fuel flow through the jet. The idle-speed adjusting screw bears against the throttle stop to regulate idle rpm.

All Briggs carburetors have an idle speed adjusting screw. Most of the newer types dispense with one or both of the mixture-adjustment screws, as shown in Fig. 4-1A. Fixed-main and/or low-speed jets should require no attention unless the calibration is upset by a change in altitude. An engine set up by the factory for sea-level operation will run rich in the rarefied air at high altitudes. Briggs can supply the correct jetting. A field fix that works on some B & S-supplied (and other) Walbro carburetors is to remove the air-bleed jet.

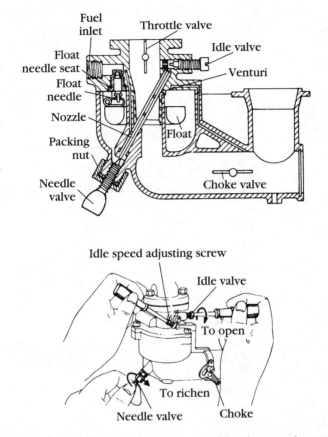

4-4 *Two-piece Flo-Jet in sectional view (A). The part that Briggs calls a "needle valve" is better known as the main, or high-speed, mixture-control screw. The "idle valve" is the idle, or low-speed, mixture-control screw. The locations of these two mixture-control screws varies with carburetor type but the idle-mixture screw is always the one closer to the engine. Because these screws control fuel delivery, backing them out of their jets enriches the mixture (B). Occasionally, you might encounter a foreign or vintage American carburetor that employs adjustable air jets, distinguished by the rounded needle tips of the adjustment screws. Backing out an air screw leans the mixture.* Briggs & Stratton Corp.

Initial mixture screw adjustment

As a rule of thumb, engines should start when the mixture-control screws are backed out 1¼ to 1½ turns from lightly seated.

Caution: Use your thumb and index finger to seat the screws. Do not force the issue with a screwdriver. Adjustment screws that have been damaged by overtightening must be replaced if the engine is to run properly (Fig. 4-5).

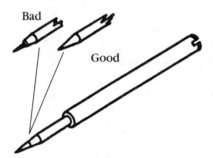

4-5 *Bent or grooved ad-justment needles must be replaced.* Briggs & Stratton Corp.

Procedure

With a clean air filter in place, fully open choke, and fresh fuel in the tank, run the engine under moderate throttle for about five minutes to reach operating temperature. The tank should be about half full on engines equipped with Vacu-Jet carburetors to minimize the effects of fuel level on mixture strength. This requirement does not apply to other carburetors.

1. Run the engine at about three-quarters speed.
2. Back out the main mixture-control screw in small increments—no more than an eighth of a turn at a time. Pause after each adjustment for the effect to be felt. Stop when engine rpm drops and, using the screwdriver slot as reference, note the position of the screw at the rich limit.
3. Tighten the screw in increments as before. Stop when engine speed falters at the onset of lean roll, which represents the leanest mixture that supports combustion.
4. Open the adjustment screw to the midpoint between the onset of lean roll and the rich limit.
5. Close the throttle and adjust the idle mixture for the fastest idle. You need at least 1700 rpm.
6. Snap the throttle butterfly open with your finger. If the engine hesitates, back out the high-speed adjustment screw a sixteenth of a turn or so and repeat the experiment. Enriching the high-speed mixture usually calls for a slightly leaner idle mixture.

The adjustment always imposes some compromise between idle quality and high-speed responsiveness. This is especially true for Vacu- and Pulsa-Jet carburetors that regulate both mixtures with a single adjustment screw. Always err on the side of richness, and do not consider any carburetor adjustment final until proven under load.

Troubleshooting

Make the checks described in chapter 2 before assuming something has gone amiss with the fuel system. Of course, sludge in the tank or raw gasoline dribbling from the air horn are powerful arguments for immediate action.

No fuel delivery

The engine appears to develop compression and the carburetor is not obviously loose on its mountings. The spark-plug tip remains dry after prolonged cranking. When mixture screws are present, backing them out has no effect. An injection of carburetor cleaner through the spark-plug port brings the engine back to life, but only for a few seconds.

Tank-mounted, Vacu-Jet carburetors suffer total failure when the check ball in the fuel pipe sticks in the closed position, almost always as a result of stale gasoline. Its cousin, the Pulsa-Jet, quits when its fuel pump diaphragm stretches or ruptures. The high- and low-speed jets (actually discharge ports) in these carburetors do not often clog and hardly ever do so simultaneously.

Float-type carburetors are susceptible to blockages between the tank and carburetor-inlet fitting. Possibilities include a clogged tank screen, fuel-tank cutoff valve, or filter on engines so equipped. External and crossover Flo-Jet fuel pumps might also fail because of an internal malfunction or a vacuum leak. (Older engines sometimes used mechanical pumps, which are susceptible to diaphragm and check-valve failure.) Check fuel delivery by cracking the line at the carburetor inlet.

If no fuel is present, work backwards, connection by connection, to the tank.

Warning: Opening fuel lines is always hazardous and especially so if the engine must be cranked to activate a fuel pump. Make these determinations outdoors with the ignition switched off and the spark-plug lead solidly grounded.

If fuel appears at the carburetor inlet, it should also be inside the instrument. Remove the fuel bowl, which is secured by a central nut on the underside. The bowl should be full. If not, the problem is a stuck inlet needle, hung float, or clogged inlet screen (on units with this feature). Failure to transfer fuel out of the bowl suggests a clogged main jet, fuel-delivery nozzle, or loss of manifold vacuum.

Engine runs lean at full throttle

This fault will appear as loss of power, possible backfire as the throttle plate is suddenly opened, and a dead-white or bleached-brown spark-plug tip. The engine seems to run better when choked. Backing out the high-speed adjustment screw (when present) has no effect.

Begin by looking for air leaks downstream of the throttle plate. Focus on the carburetor mounting flange and cylinder head gasket. Lean running in a two-cycle engine is the classic symptom of crankshaft seal failure, but don't settle on this rather grim diagnosis until other possibilities have been eliminated.

Replace the optional fuel filter and open the line to verify that copious amounts of fuel are available at the carburetor inlet fitting. Note the preceding warning about ignition sparks and spilled gasoline. Finally, look for a stoppage in the high-speed circuit, which will usually be at the point of discharge.

Engine runs rich

A blackened spark plug, acrid, smoky exhaust, and loss of power suggest an overly rich mixture. It is assumed that turning the mixture adjustment screws (when present) has no effect and that the air filter has been cleaned or, if made of paper, replaced. The choke butterfly opens fully, as verified by visual inspection with the air cleaner removed.

In my experience, persistently rich mixtures are a problem almost entirely confined to float-type carburetors. Replace the needle and seat and set the float level to specification. If the difficulty persists, check for clogged air bleeds. Briggs & Stratton Walbro carburetors use a replaceable high-speed air jet that might have been removed in an ill-advised attempt to richen the mixture.

Note: Fixed-jet carburetors require recalibration at high altitudes. Contact your dealer for recommended fuel and air jet sizes.

Carburetor floods

Long bouts of cranking with the choke closed will flood any carburetor, wet the bore, and spill fuel out of the air horn. Two-piece Flo-Jets flood quite easily.

Note: Carburetor flooding, characterized by fuel puddling in the bore, must be distinguished from external leaks. Fuel will cascade past a worn or twisted float-bowl gasket. Tank-mounted Vacu- and Pulsa-Jet carburetors might weep fuel at the tank interface because of cracks in the tank or a bad gasket. See the following "Removal and installation" section for additional information.

Failure of the inlet needle and seat or of the float mechanism produces spontaneous flooding in Flo-Jet and B & S Walbro car-buretors. Needle-and-seat failures are usually attributable to wear, although dirt in the fuel supply can produce the same effect. Dirt-induced flooding might spontaneously cure itself, only to reappear as another particle becomes trapped between the needle and seat. Clean the fuel system and replace the needle and seat.

Float failures are usually of the obvious mechanical sort and correctable by cleaning.

Engine refuses to idle

Engines that operate under a constant load regime might not have provision for idle. Once the engine starts, the governor raises engine speed to a preset rpm. This discussion applies to engines that left the factory with an idle capability and now refuse to exercise it.

Check for air leaks at the carburetor mounting flange, cylinder-head gasket, and at the throttle-shaft pivots. The latter source might not be significant but will allow abrasives to enter the engine. Other possibilities include:

- *Idle rpm set too high.* All Briggs & Stratton carburetors have an adjustable throttle stop in the form of a spring-loaded screw. Use a tachometer to adjust to the factory or equipment manufacturer's specification.
  Caution: Air-cooled, splash-lubricated engines do not idle in the automotive sense of the word. Speeds of 1700 rpm and more are the norm.
- *Maladjusted throttle cable.* Loosen the cable anchor and reposition the Bowden cable as necessary.
- *Binding throttle shaft or linkage.* New throttle shafts sometimes bind because of paint accumulations. Grass or other debris might limit the freedom of movement of the throttle-return mechanism.
- *Governor failure.* With the engine running and the throttle lever set on idle, gently try to close the throttle. Do not force the issue. If light finger pressure does not move the

throttle against its stop, a possible governor malfunction is indicated for engines intended to run at variable speeds.

• *Clogged low-speed circuit.* Clean the carburetor.

Removal & installation

The carburetor flanges to the engine block or makes a slip-fit connection, sealed by an O-ring, with the fuel-inlet pipe. The fuel supply must be shut off on gravity-fed systems, either with the tap provided or by inserting a plug into the carburetor end of the flexible fuel hose.

Warning: Some gasoline will be spilled. Work outside in an area remote from possible ignition sources.

The governor mechanism must be disengaged from the throttle arm without doing violence to the associated springs and wire links. Some springs have open-ended loops and can be easily disengaged with long-nosed pliers. Others incorporate double-ended loops that come off and go on in a manner reminiscent of the "twisted-nail" puzzle.

Note the lay of the spring and, if there is any possibility of confusion, mark the attachment holes.

Wire links remain connected until the carburetor is detached from the engine. While holding the carburetor in one hand, twist and rotate it out of engagement with the links, being careful not to bend the wires in the process.

Four, and on one model five, screws secure Vacu- and Pulsa-Jet carburetors to the fuel tank. Automatic-choke models used on 920000, 940000, 110900, and 111900 engines have one of the screws hidden under the choke butterfly. Inspect the tank interface for cracks (in which case, the tank must be replaced) and for low spots that might cause fuel or vacuum leaks (Fig. 4-6). This problem is serious enough for the factory to supply a Pulsa-Jet tank repair kit (PN 391413).

Assembly is the reverse of disassembly. Always use new gaskets mated against clean flange surfaces.

Warning: Briggs & Stratton engines built before the early 1970s (the factory spokesperson contacted could not be more specific about the date) used asbestos gaskets. Soak the area with oil and remove gasket material with a single-edged razor blade and dispose of the shards safely. Do not use a wire wheel or any other method that might create dust.

Make up the links first, then secure the carburetor to the engine. Connect the springs last.

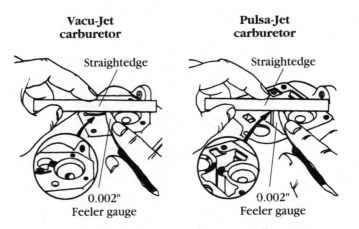

4-6 *Vacu-Jet and Pulsa-Jet tank flanges are crucial. Hairline cracks or low spots leak fuel. A low spot in the shaded area shown (B) denies vacuum to the Pulsa-Jet fuel pump.* Briggs & Stratton Corp.

Automatic choke   Most late-production Vacu- and Pulsa-Jet carburetors are equipped with a type of automatic choke unique to Briggs & Stratton. The choke valve is normally closed by a spring and pulled open by a manifold vacuum acting on a large diaphragm.

The choke butterfly should snap closed when the engine stops and open as soon as it starts. If should also flutter in response to changes in the manifold vacuum under severe loads. In this function, the automatic choke acts as an enrichment device.

Spring length is critical:

Application	Spring free length
Vacu-Jet	$^{15}\!/_{16}$ to 1 in.
Pulsa-Jet	$1\frac{1}{8}$ to $1\frac{7}{32}$ in.
Engine models 11090 & 11190	$1^{15}\!/_{16}$ to $1\frac{3}{8}$ in.

To assemble:

1. If a new diaphragm is being installed, attach the spring as shown in Fig. 4-7.
2. Invert the carburetor and guide the spring/link assembly into its recess (Fig. 4-8).
3. Turn the assembly over and start the mounting screws, turning them just far enough for purchase.

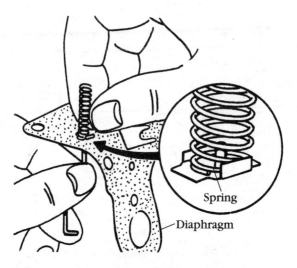

4-7 *Attach the choke link to the diaphragm as shown.* Briggs & Stratton Corp.

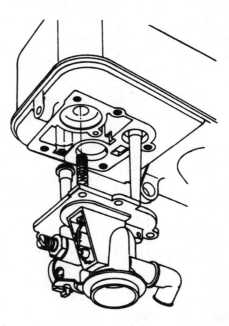

4-8 *With the parts inverted, mate the carburetor to the tank. Turn the assembly over and lightly start the hold-down screws. The parts must be free to move relative to each other.* Briggs & Stratton Corp.

4. Depress the choke butterfly with one finger and attach the actuating link (Fig. 4-9). This action preloads the diaphragm spring.

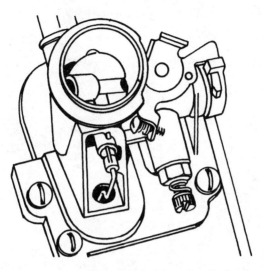

4-9 *Complete the assembly while holding the choke closed.* Briggs & Stratton Corp.

5. Tighten the mounting screws, running them down incrementally in an X-pattern. Spring preload should lightly tip the choke over center, toward the closed position. Install the link cover and gasket.

If the choke remains closed after the engine starts, the problem can usually be traced to insufficient spring preload. Too much spring preload will hold the choke open, regardless of the strength of the vacuum signal. Fuel or oil in the diaphragm chamber has the same effect. Other possibilities include carbon/varnish deposits on the butterfly pivots or a bent air-cleaner stud. If the stud is bent, replace it, rather than attempting a repair that might compromise the air cleaner-to-carburetor seal.

Pulsa-Jet tank diaphragm Pulsa-Jet carburetors include a fuel pump, which is actuated by a vacuum diaphragm. Side-mounted diaphragms are discussed in an upcoming section titled "Vacu-Jet & Pulsa-Jet." The tank-mounted diaphragm (Fig. 4-10), used on one version of this carburetor, concerns us here.

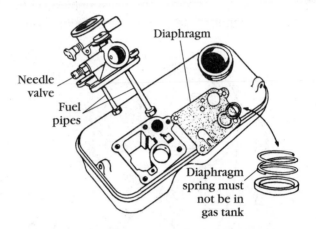

4-10 *Pulsa-Jet diaphragms install on the tank followed by the wear collar and spring.* Briggs & Stratton Corp.

Figure 4-6 illustrates the places where vacuum leaks typically develop at the tank interface. Figure 4-10 shows the sequence of assembly. Note that the spring rests collar down on the top of the diaphragm and not in the tank cavity.

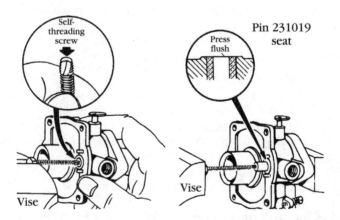

4-11 *Briggs & Stratton uses elastomer inlet seats, which are extracted with a self-tapping screw and pressed into place using the original seat as a buffer. In the case of the one-piece Flo-Jet, the seat should be flush with the raised casting lip.*

Repair & cleaning

Carburetors do not wear out in the accepted sense of the term. Most failures are associated with "soft" parts, such as inlet needles and seats, diaphragms, and gaskets. After long service, the throttle shaft bearings might develop enough play to justify replacement, when such repairs are possible.

Light varnish deposits respond to lacquer thinner and compressed air. Heavier deposits can be removed with Gunk Carburetor Cleaner, which as these products go, appears to be fairly benign. But no chemical cleaner, however aggressive, can undo the effects of water-induced corrosion. White powdery deposits and leached castings mean that the carburetor should be discarded. Sometimes you can get a water-damaged carburetor to work, but it will never be quite right.

Strip off nonmetallic parts, including gaskets, O-rings, elastomer inlet seats, and diaphragms. Leave nylon butterflies in place, since a short exposure to Gunk does not seem to hurt. Nylon-bodied carburetors can also be cleaned, provided immersion is limited to five minutes or so.

As you clean and repair, don't disturb the following items:

- The throttle and choke butterfly valves.
- The Vacu- and Pulsa-Jet pickup tubes (unless for replacement).
- Any part that resists ordinary methods of disassembly. Threaded, as opposed to permanently installed pressed-fitted parts, have provisions for screwdriver or wrench purchase. But forced removal might do more damage than carburetor cleaner can correct.
- Expansion plugs (unless loose or leaking). Briggs & Stratton carburetor overhaul kits include expansion plugs. If you opt to replace these items, install them with a flat-nosed punch (a piece of wood will do) sized to the plug's outside diameter (OD). Seal by running a bead of fingernail polish over the joint after installation.

Carburetor service by model

Depending on displacement, market, and vintage, Briggs & Stratton engines employ any of five basic carburetor types. Each type includes subvariants and most have undergone running production changes, which are keyed to the engine build date.

Two-piece Flo-Jet

Small, medium, and large two-piece Flo-Jets are the only up-draft carburetors in the Briggs line. Figure 4-4A shows the parts layout common to all three models.

Needle & seat   Extract the elastomer seat by threading a self-tapping screw into the fuel orifice (Fig. 4-11). Press in a replacement seat—PN 230996 for gravity feed, PN 231019 for applications with a fuel pump—flush with the casting. Viton-tipped needles can be reused, if not obviously worn.

Float setting   When inverted and assembled without a bowl-cover gasket, the float should rest level with the casting (Fig. 4-12). Adjust float height by bending the tang with long-nosed pliers. Figure 4-13 illustrates the proper orientation of the needle clip for this and other carburetors that employ the device.

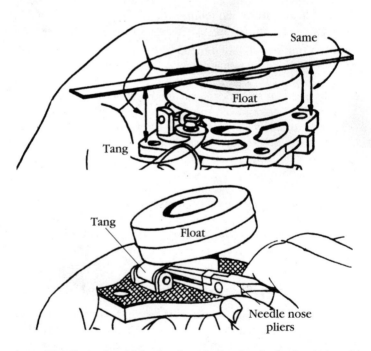

4-12 *The float should be level with the (gasketless) casting for Flo-Jet and all other Briggs carburetors. Adjust by bending the tang without applying force to the needle.*

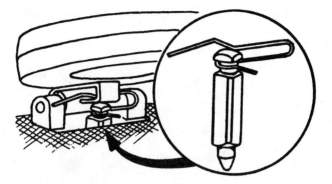

4-13 *Install needle spring as shown for all applications.* Briggs & Stratton Corp.

Caution: Do not make the float height adjustment by pressing the float against the needle.

Casting distortion Overtightening the hold-down screws distorts the bowl-cover casting. Assemble without a gasket, and check with a 0.002-in. feeler gauge (Fig. 4-14). If the blade enters, remove the cover and straighten the "ears" with light hammer taps.

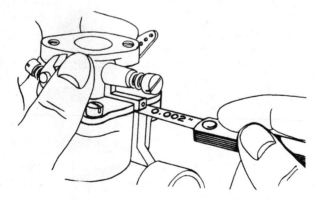

4-14 *The two-piece Flo-Jet bowl-cover casting (or "throttle body") is a fragile casting, easily warped by over tightening. Allowable distortion is <0.002 in., as measured with a 0.5-in.-wide feeler gauge. If the gap is excessive, turn the casting over and, using a small hammer, tap the corners back into alignment.* Briggs & Stratton Corp.

Throttle shaft/bearing replacement   Follow this procedure:

1. Using a ⅛-in. punch, drive out the roll pin that secures the throttle lever to its shaft.
2. Scribe marks on the throttle blade and carburetor body as assembly references.
3. Remove the two small screws that secure the throttle butterfly to the shaft.
4. Remove the butterfly and shaft.
5. Extract the shaft bushings with a ¼-in. tap.
6. Press in new bushings to original depth.
7. Install a new throttle shaft and throttle butterfly with scribe marks indexed. Coat the butterfly screw threads with Loc-tite. Start the screws, but do not tighten.
8. Close the butterfly to center it in the carburetor bore and tighten the screws. Verify that the butterfly swings through its full arc without interference. Complete the assembly.

One-piece Flo-Jet

Briggs manufactures one-piece Flo-Jets in two sizes. The small version carries its main jet in the float bowl (Fig. 4-15A). The large model employs a remotely located main jet, supplied through a removable nozzle (Fig. 4-15B).

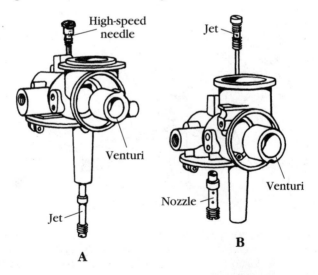

4-15 *Small (A) and large (B) one-piece Flo-Jets with float bowls removed. The large model has its high-speed mixture screw under the float bowl.* Briggs & Stratton Corp.

Needle & seat Service as described for the two-piece Flo-Jet (Fig. 4-11). Note that the replacement seat must be dead flush with the casting.

Float setting Adjust as described for the two-piece Flo-Jet. When inverted, the float should rest level with the carburetor body (Fig. 4-12).

Crossover Flo-Jet

Figure 4-16 is a sectional view of the crossover Flo-Jet, as used on horizontal-crankshaft Model 253400 and 255400 engines. This carburetor includes a vacuum-operated fuel pump, illustrated in the next drawing.

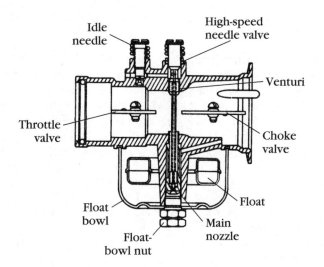

4-16 *Crossover one-piece Flo-Jet adjusts from the top.* Briggs & Stratton Corp.

Needle and seat renewal and float adjustment procedures are as described for the two-piece Flo-Jet. Figure 4-17 illustrates the assembly sequence for the double-diaphragm fuel pump, normally serviced with a rebuild kit that contains springs and "soft" parts.

Briggs & Stratton Walbro

The factory appears to be slowly phasing out the Flo-Jet series in favor of highly modified Walbro carburetors. Engines in the 9-

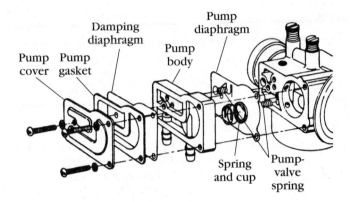

4-17 *Briggs supplies replacement diaphragms singly or as part of the crossover Flo-Jet rebuild kit. Assemble dry, without sealant.*

to 13-cubic-inch range use variants of the small series Walbro, recognized by its angular appearance and removable air bleed jet, mounted just aft of the air cleaner. The removable main jet nozzle has been omitted, together with replaceable throttle-shaft bushings and other niceties associated with earlier designs. Most small Walbros have a fixed-main jet and an adjustable low-speed jet, but there are exceptions. The two-cycle carburetor, pictured in Fig. 4-18, employs fixed jets for both circuits. Carburetors used on 120000 engines are completely adjustable.

The large B & S Walbro, fitted to 19-, 25-, and 28-CID vertical shaft engines, employs an external air bleed jet, which is located next to the idle mixture screw, and a removable main nozzle, which is accessed from the float bowl. Fixed-main jets are the norm.

Aside from the nozzle detail, service procedures are similar for the small and large versions of the carburetor.

Inlet seat Fish out the elastomer seat with a hooked wire and install the replacement to cavity depth (Fig. 4-19) using a flat punch sized to seat OD.

Float Float pins might include an antirotation feature in the form of flats milled on one end. Drive out the pin from the unmarked end and assemble with the pin flats aligned to corresponding flats on the pivot bearing (Fig. 4-20). A spring clip ties the needle to the float (Fig. 4-21). Float height is fixed by needle and seat geometry and should not be tampered with in the field.

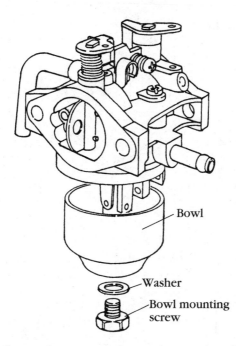

Bowl

Washer

Bowl mounting screw

4-18 *Walbro carburetor, used on 95700 and 95600 two-cycle engines, employs a fixed jetting. The L-shaped tube on the left vents the float bowl.* Briggs & Stratton Corp.

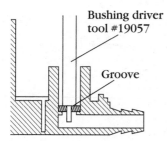

Bushing driver tool #19057

Groove

4-19 *Press the fuel-inlet seat home with the grooved side of the seat down, toward the in-coming fuel stream. A flat punch, sized to seat OD, can be used in lieu of the factory tool shown.*

Vacu-Jet & Pulsa-Jet

Figure 4-2 illustrates the Vacu-Jet mechanism, which is distinguished by a single pickup tube and tandem discharge ports controlled by flow through a single jet. Figure 4-22 shows the three basic forms of this carburetor.

The Pulsa-Jet derives from the Vacu-Jet and in its various permutations uses many of the same parts. The distinction be-

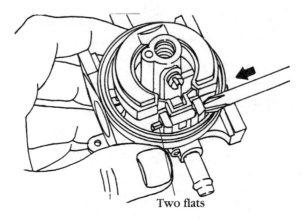

Two flats

4-20 *Flats are provided on some float pins, apparently to prevent hanger wear. Extract from the flattened end and install with flats properly indexed.* Briggs & Stratton Corp.

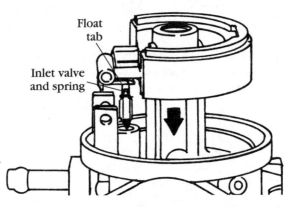

Float tab

Inlet valve and spring

4-21 *Install the float with the needle assembled to its clip. Do not tamper with the float tang.* Briggs & Stratton Corp.

tween the two is that the Pulsa-Jet feeds from a reservoir in the top of the fuel tank, which it replenishes with a vacuum-powered fuel pump (Fig. 4-23). Pulsa-Jets have two pickup tubes. The longer one transfers fuel from the tank to the reservoir; the shorter tube draws from the reservoir into the carburetor. This arrangement isolates the carburetor from changes in the level of fuel in the tank. Vacu-Jets lean out as the tank depletes.

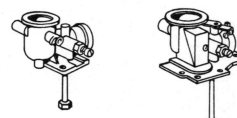

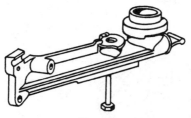

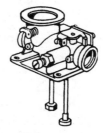

Vacu-Jet

A

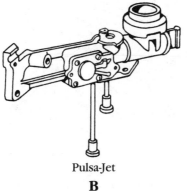

Pulsa-Jet

B

4-22 *Vacu-Jet (A) and Pulsa-Jet (B) variations.* Briggs & Stratton Corp.

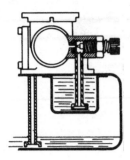

4-23 *Three pulls of the starter cord should pump enough fuel into the Pulsa-Jet reservoir to start the engine. Once it starts, the reservoir remains full to a level defined by a spill port. Thus, the internal fuel level of the carburetor is independent of the level of fuel in the tank. According to Briggs & Stratton, a Pulsa-Jet-equipped engine develops as much power as one supplied by a more expensive float-type carburetor.*

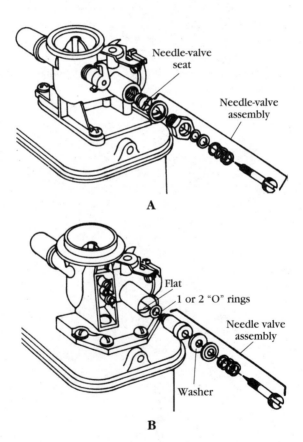

4-24 *Two needle valve assemblies are used, one primarily associated with zinc carburetors (A); the other is used on all nylon-bodied models (B).* Briggs & Stratton Corp.

Figure 4-24 illustrates major Pulsa-Jet variations that closely track those of the Vacu-Jet. Most service information applies to both types.

Needle-valve assembly   Sealing the needle valve, or mixture-adjustment screw, involves some fairly complex engineering. Figure 4-25A shows the arrangement of washers and O-rings generally found on pot-metal Vacu- and Pulsa-Jets. Figure 4-25B illustrates the arrangement always used on nylon carburetors and sometimes on the zinc models. The needle is quite vulnerable to damage from overtightening.

Pickup tubes   Vacu-Jet fuel pickup tubes are fitted with a check ball, which tends to stick in the closed position.

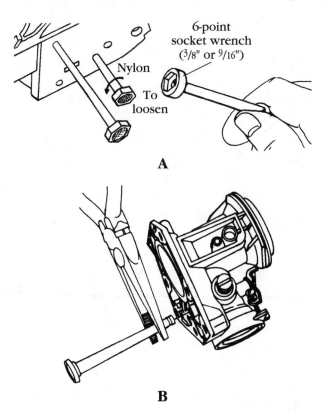

4-25 *Fuel pipes twist in and out of metallic carburetor castings (A). Pipes for nylon carburetors incorporate a snap lock (B).* Briggs & Stratton Corp.

Because the ball and, on later models, the tube itself are made of nylon, more than a few minutes in carburetor cleaner is all that can be tolerated. As an emergency repair, you can free the ball by gently inserting a fine wire through the screen in the base of the tube. Eventually the assembly will have to be replaced.

Fuel pickup tubes supplied with zinc carburetors twist off and on. Tubes used with nylon carburetors snap in and out, an operation that can require considerable force (Fig. 4-25).

Pulsa-Jet pump diaphragm   The side-mounted diaphragm is shown in Fig. 4-26. The tank-mounted version, used with "bobtail" carburetors, is illustrated in Fig. 4-8. In either case, replace the diaphragm whenever the carburetor is serviced.

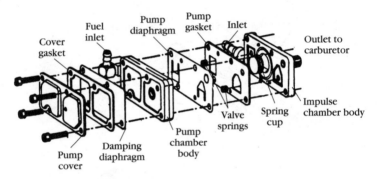

4-26 *Diaphragm-type fuel pump.* Briggs & Stratton Corp.

Fuel pump

Some engines are equipped with stand-alone versions of the crossover Flo-Jet pump. Fuel enters by gravity and leaves under pressure generated by a pulse-activated diaphragm. Replace the pump diaphragm as a routine part of fuel-system service.

Air filters

Briggs & Stratton supplies a variety of standard and optional air filters, which use foam or paper elements in combination with foam precleaners or "socks (Fig. 4-27)." The foam in early pro-

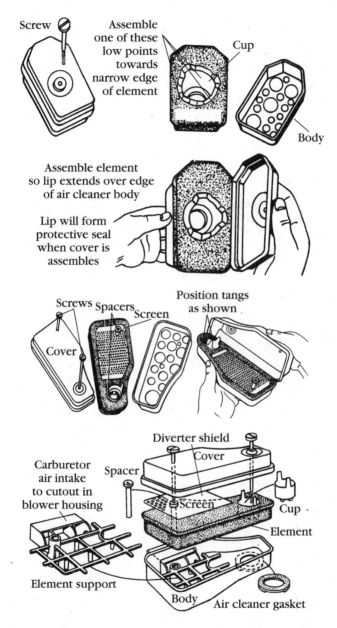

4-27 *Three Briggs foam-type air cleaners, illustrated to show correct assembly procedures, which might not be obvious to the average mechanic. The more sophisticated paper-and-foam units offer less opportunity for error.*

duction filters tolerated kerosene and other petroleum-based solvents; the foam used on later designs does not. Wash all filter elements in warm water and nonsudsing detergent to avoid confusion and ruined filters. Rinse in the reverse direction of air flow until the water runs clear. Towel off the excess water and oil the element, kneading it as shown in Fig. 4-28. Foam elements should be re-oiled as needed and whenever an engine that has been out of service for more than a few days is started. Oil tends to migrate out of the foam.

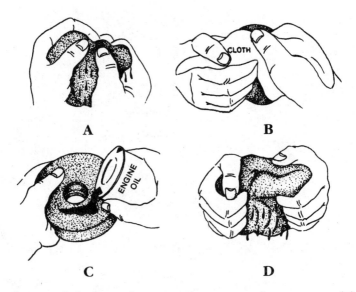

4-28 *Wash the foam element in detergent and warm water (A). Wrap with a shop towel and press dry (B). Oil the element, wetting it thoroughly (C), and squeeze out the excess (D). Don't over-oil precleaners that stand against paper elements.* Briggs & Stratton Corp.

Pleated-paper elements, of whatever vintage, must not be exposed to oil or petroleum solvents. Wash paper elements in warm water and nonsudsing detergent. Allow the element to dry and the pores to shrink before installing. Do not blow out these fragile elements with compressed air.

The cleaner-to-carburetor gasket should be renewed at the first sign of wear. Older and less expensive engines secure the filter with a single screw that, if bent, destroys the gasket seal. Replace the screw as necessary.

Governors

Small engine governors put a ceiling on no-load speed and hold rpm relatively constant under varying loads. Less expensive engines typically use air-vane governors of the general type illustrated in Fig. 4-29. The spring tends to close the throttle; the dynamic head of cooling air acting against the vane attempts to open the throttle. The manually operated throttle varies spring tension, relaxing it to allow the engine to run faster or, moved in the other direction, stretching the spring to slow the engine.

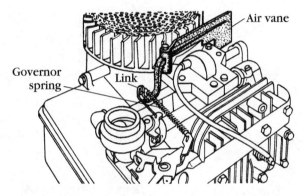

4-29 *The typical Briggs & Stratton governor uses a plastic vane loosely secured with metal tabs.*

Centrifugal governors work in the same manner except that the closing force is generated by spinning weights, called *fly weights* (Figs. 4-30 and 4-31). These governors can be quite complex in detail and resist generalization.

Most governor malfunctions—hunting, lack of responsiveness, excessive no-load speed—can be corrected by replacing weak or distorted throttle springs and/or bent wire links. Do not change the geometry by installing springs or links in alternate mounting holes that might be present.

Warning: Governor springs are safety items that exert major influence on no-load speed. Ungoverned engines can act like grenades, exploding in fragments. Replace springs with the correct part number. Do not stretch or distort during installation. After any spring change, check no-load speed

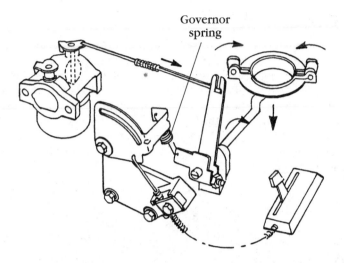

4-30 *Centrifugal governor operation in response to load. Engine speed drops and the weights slow, allowing the spring to pull the throttle open wider. Details vary with engine model, but the pattern of forces—the spring pulls the throttle open, the weights tend to close it with a force proportional to engine rpm—applies to all.* Briggs & Stratton Corp.

against the equipment manufacturer's specification with an accurate tachometer.

Mechanical governors incorporate an adjustment that, wrongly accomplished, affords the unwitting mechanic the opportunity to destroy the engine within seconds of startup. In most cases, the adjustment involves loosening the pinch bolt (Fig. 4-31) and rotating the shaft in a specified direction, relative to the lever. Because of the critical nature of this work, it should be farmed out to your local Briggs & Stratton distributor who, presumably, will stand good for mistakes.

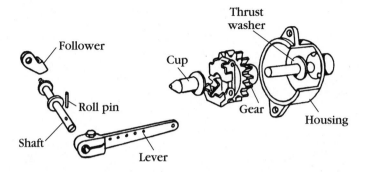

4-31 *The governor mechanism used for 60000, 80000, and 140000 engines. The housing, accessed from outside of the engine, is unique to this engine family; all other centrifugal governors live inside of the crankcase. The pinch bolt that secures the lever and shaft is the main adjustment point for this and most other Briggs & Stratton governors and **should not be disturbed during normal service activities, including engine overhaul.***

5

Starters

Current Briggs & Stratton engines use a side-pull or vertical-pull rewind starter that might be supplemented with a dc- or ac-starter motor. Formerly, some of the company's products were fitted with a spring-powered unit, which is discussed briefly at the end of this section.

Rewind starters

The *rewind starter*, also known as the *recoil starter*, was introduced by Jacobsen in 1928 and has since become standard for small engines. While the constructional details differ, rewind starters have the following basic components:

- Pressed steel or aluminum housing, which locates the starter center over the flywheel center.
- Recoil spring, one end of which is anchored to the housing; the other to the pulley. The spring mounts in a recess in the pulley or housing.
- Nylon starter cord, sized by diameter and length to the application and secured to the pulley at one end and the handle at the other.
- Pulley, also known as the sheave ("shiv"), located relative to the starter housing by tabs or a bushing.
- Clutch, which transfers starter torque to the flywheel and automatically disengages when the engine starts. Briggs starters use either a sprag or friction clutch.

Troubleshooting
This section details some of the more common failures.

Broken rope This is the most common failure and is often caused by the operator pulling too hard at the end of travel. Rope breakage might be abetted by a worn guide bushing (the ferrule that protects the rope from contact with the housing) or by a tendency of the engine to kick back during cranking.

Rope won't completely retract Some spring action is present, which might be frustrated by starter-to-flywheel misalignment or by loss of preload. With the rope extended and the blower-housing bolts loosened a few turns, strike the housing with the palm of your hand. If the rope retracts, misalignment was the problem and the housing can be secured. If repositioning the housing has no effect, the recoil spring should be replaced. It is possible to salvage tired springs by increasing the preload, but at the cost of reduced rope travel.

No spring action If the rope goes limp, the cause is almost certainly a broken spring.

Rope is hard to pull Check the starter-to-flywheel alignment as described previously. Other possibilities include abnormal engine friction or, if rope resistance seems to pulsate, a loose flywheel or rotary mower blade. Check the sheave-axle bearing on Eaton starters.

Starter slips, and fails to engage the flywheel This problem originates in the starter clutch and, on Eaton-pattern starters, almost certainly indicates a failed brake spring. Briggs clutches slip if extremely dirty.

Starter howls as engine runs This is the classic symptom of a dry clutch/crankshaft bearing on Briggs-pattern starters. Remove the blower housing and apply a few drops of oil to the crankshaft end.

Eaton-pattern, side-pull starter

Quantum, Europa, and the better engines generally use versions of the Eaton starter. An Eaton starter employs spring-loaded clutch dogs that cam into engagement against the flywheel hub inside diameter (ID). The starter assembly is generally secured to the blower housing with pop rivets, although some engine models use bolts (Fig. 5-1). In any event, replacement starter assemblies bolt into place using the existing mounting holes.

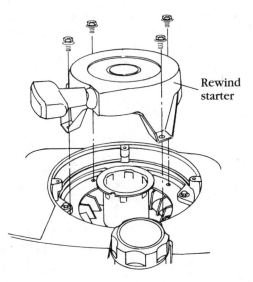

Rewind
starter

5-1 *We will be seeing more Eaton starters as Briggs turns to outside vendors for components. While these starters are not identical to those used by other manufacturers, clutch parts can be purchased from Tecumseh dealers. This illustration does not show the plastic grass guard that mounts over the starter housing.*

Disarming

First release spring preload tension, which on Eaton starters is done by removing the handle and allowing the sheave to unwind in a controlled fashion. Brake the sheave with your thumbs. It is also helpful to count the number of sheave rotations from the point of full-rope retraction so that the same preload can be established upon assembly.

Warning: Even after preload has been dissipated, the spring remains confined in its housing under considerable tension. Wear safety glasses when servicing these starters.

Rope replacement

If the rope has parted, preload has already been lost. If the rope remains in one piece, untie the knot securing the handle. Allow the rope to fully retract, braking the sheave as described in the preceding paragraph.

Briggs & Stratton supplies precut and fused starter cords for the various engine models. If you purchase stranded nylon cord in bulk, replicate the original diameter (#4½ for 60000 through 120000; #5½ for 130000 and larger) and length. Fuse the cut ends in an open flame.

Warning: Fused rope ends retain enough heat to produce painful burns for several minutes after exposure to flame.

Follow this procedure:

1. Tie a square knot in one end of the rope (Fig. 5-2).

5-2 *A figure-eight knot secures the rope to the sheave. Rope ends should be melted to prevent unraveling.* Briggs & Stratton Corp.

2. Using a screwdriver in the sheave slot, wind the spring clockwise (as viewed from the underside of the starter) until tight.
3. Allow the sheave to unwind enough to align the slot with the rope eyelet (Fig. 5-3).

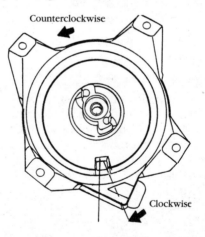

Counterclockwise

Clockwise

5-3 *Turn the sheave counterclockwise to coil bind and back off far enough to align the rope cavity with the eyelet. This establishes preload for the Briggs-Eaton starters discussed here. If you are dealing with an unfamiliar starter, preload is the number of revolutions the sheave made after "swallowing" the rope. If, after assembly, the rope fails to retract smartly, add a revolution or so of preload. If the spring coil binds near the end of rope travel, release some of the preload.*

4. Insert the unknotted end of the rope into the slot and through the eyelet (Fig. 5-4).

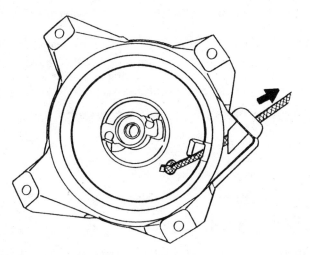

5-4 *Thread the rope through the eyelet, or bushing, and seat the knot in the sheave cavity. To avoid fighting the spring while tying on the handle, clamp the sheave to the housing with Vise-Grips.* Briggs & Stratton Corp.

5. Tie a temporary knot in the rope or lock the sheave with Vise-Grip pliers (Fig. 5-5).
6. Install the handle and secure with a knot. Release the sheave and test starter action.

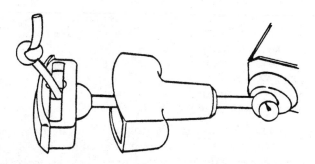

5-5 *Knot the rope as shown at the handle.* Briggs & Stratton Corp.

Clutch

Two spring-loaded dogs inside a pressed aluminum retainer transmit starting torque to the flywheel hub. A compression pin or screw secures the retainer to the underside of the sheave and the sheave to its axle. Note that the two-cycle retainer screw has left-hand threads.

Warning: Wear eye protection when servicing rewind starters, particularly during and subsequent to removal of the retainer fastener. All that holds the main spring captive is a shallow recess in the upper side of the sheave.

Engine model	Clutch retainer fastener	Torque
95700, 96700 (2-cyc)	Hex-head, LH thread	30 lb./in.
99700 (Europa)	Pin	
104700 (OHV)	Phillips, RH thread	70 lb./in.
120000 (Quantum)	Pin	

Figures 5-6A and 5-6B illustrate these arrangements. Note that the fastener preloads the brake spring against the retainer. Spring tension causes the retainer and sheave to turn together as an assembly during the first few degrees of sheave rotation. Further movement of the retainer is then blocked by the dogs that, at this point, grip the flywheel hub in full extension. Continued rotation of the sheave exerts a steady drag on the retainer to hold the dogs in engagement. When the starter cord is released, the sheave rewinds and drags the retainer with it for a fraction of a turn, just enough to back out the dogs and release the flywheel. In practical translation, this means:

- Brake springs are the most vulnerable part of the assembly.
- A loose retainer pin or screw dissipates brake spring preload, resulting in starter slip.
- Retainers undergo severe wear at the brake spring and dog contact points.
- Bent or distorted dog springs should be replaced.
- Lubrication is an enemy.

Use a punch to drive out the compression pin from the outboard side of the starter housing to the engine side. In some cases, a decorative decal must be peeled from the starter hous-

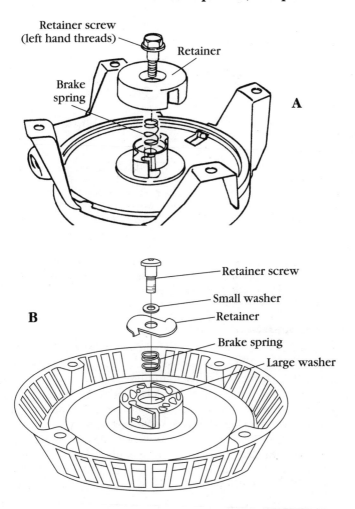

5-6 *Eaton starters are held together with a screw (A) or throwaway compression pin (B). Washer sequence varies with starter model but all employ a coil spring as a friction generator.*
Briggs & Stratton Corp.

ing to gain access to the pin. Support the housing and sheave with PN 19227 or a short length of pipe placed vertically beneath it (Fig. 5-7). From the engine side, hammer or press in a new replacement pin and install spring and washers. Seat at the original depth.

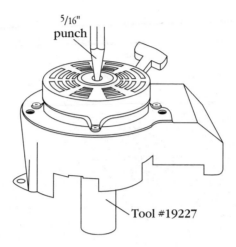

5-7 *PN 19227 is a hollow cylinder that allows the pin to drop while keeping the sheave on its axle and the main spring contained. A piece of 3-in pipe can substitute.* Briggs & Stratton Corp.

Spring & sheave

The spring lives in a recess in the sheave between the sheave and underside of the starter housing. One end of the spring anchors to the sheave, the other to the housing.

Springs used for engine Models 104700 overhead-valve (OHV) and 120000 (Quantum) are considered an integral part of the sheave and should not be separated from it.

95700 and 96700 (two-cycle) and 99700 (Europa) have replaceable springs. Grasp the spring with long-nosed pliers and carefully lift it out of the sheave (Fig. 5-8) and release it inside of the starter housing (Fig. 5-9), which acts as a kind of cage.

Warning: Wear safety glasses and long sleeves and gloves.

Inspect the sheave for cracks in the hub and damage to the spring anchor. Plastic sheaths require no lubrication, although a bit of oil on the axle helps prevent rust.

Replacement springs are packaged in a plastic retainer for ease of handling. A small dab of grease is all the lubrication required. Insert the outer spring end in the slot provided and, holding the coils with long-nosed pliers, cut the retainer loose (Fig. 5-10). Reinstalling the original spring, which does not have a retainer, is a matter of anchoring the outer end and laying the spring down counterclockwise, a coil at a time (Fig. 5-11). Install

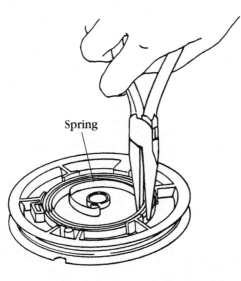

5-8 *Grasp the spring with long-nosed pliers and carefully lift it out of the sheave recess on engine models with replaceable springs.* Briggs & Stratton Corp.

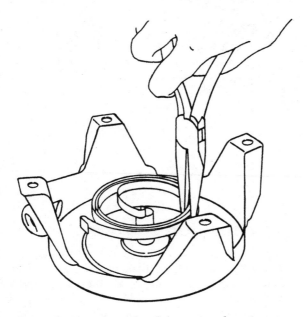

5-9 *Release the spring inside of the starter housing.* Briggs & Stratton Corp.

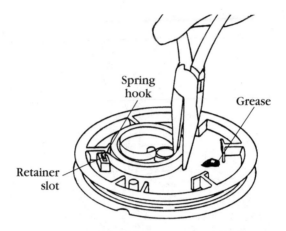

5-10 *New springs are first anchored and then released by cutting the retainer band.* Briggs & Stratton Corp.

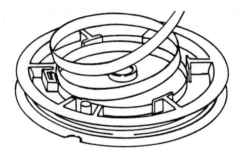

5-11 *Used springs can be wound in by hand.* Briggs & Stratton Corp.

the sheave on the axle and rotate it counterclockwise to engage the anchor.

Briggs & Stratton starters

The Briggs side-pull starter continues to be specified for most single-cylinder models. Unlike other rewind starters, it is integral with the blower housing and drives through a sprag, or rachet-type, clutch.

Sprag clutch   Recoil and impulse starters drive through a sprag clutch that doubles as the flywheel nut. The clutch housing (Fig. 5-12) threads over the crankshaft. The sprag (ratchet in the drawing) is supported by a bushing on the crankshaft stub. Its outside end mates with the starter pulley, and its lower, or inside, end rides against four or six ball bearings in the starter housing. When rotated by the starter pulley, the sprag traps a ball bearing between it and the clutch housing, locking the starter to the crankshaft. Once the engine catches, the ball bearing releases and the sprag idles on the bushing.

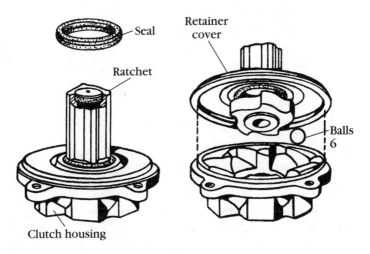

5-12 *The current production sprag clutch.* International Harvester Corp.

To service the clutch, remove the engine shroud and the screen, which is mounted to the clutch housing by four self-threading screws. Disconnect and ground the spark-plug lead to prevent accidental starting. Secure the flywheel with a strap wrench or a Briggs & Stratton holding fixture. Unthread the clutch assembly using factory tool PN 19161 or 19114. If this tool is not available, the assembly can be loosened with a hammer and a block of soft wood. Some damage to the screen lugs is inevitable but is less than fatal if distributed evenly to all four lugs. A spring washer fits under the clutch assembly.

On early models, the retainer cover was secured with a spring wire; on late models, the cover must be pried off. Clean the sprag, clutch housing, and ball bearings in solvent. Some deformation of the clutch housing is normal. Wear on the tip of the sprag, the part that makes contact with the bearings, can cause the clutch to slip. Reassemble these parts dry, without lubricant, and lightly oil the bushing. Install the spring washer and torque to specifications in Table 5-1.

Table 5-1.
Clutch housing torque limits

Cast-iron series	Torque (ft-lb)
6B, 6000, 8B, 80000, 82000, 92000, 110000	55
100000, 130000	60
140000, 170000,1717000, 190000, 191700, 251000	65
Aluminum series	
19, 190000, 200000	115

Horizontal pull starter To dismantle the starter, remove the shroud and place the assembly upside down on a bench (Fig. 5-13). Cut the rope at the sheave knot and extract it. Use a pair of pliers to pull the main spring out of the housing as far as it will come (Fig. 5-14). The purpose is to bind the spring so that it will not "explode" when the sheave is detached. For further protection, wear safety glasses. Carefully straighten the sheave tangs. Withdraw the sheave, twisting it slightly to

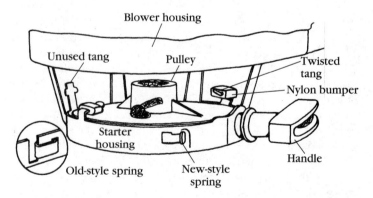

5-13 *Some models employ an insert at the housing spring-anchor slot.* Briggs & Stratton Corp.

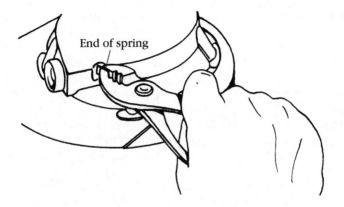

End of spring

5-14 *Disarm the spring before disassembly.* Briggs & Stratton Corp.

disengage the spring. Clean the metal parts in solvent and inspect for damage.

Secure the blower shroud to the workbench with several large nails or hold the shroud lightly in a vise. Lightly grease the spring and attach it to the sheave. Thread the free end out through the anchor slot in the shroud. The hole in the sheave measures ¾ in. square. A 6 in. length of a 1 × 1 or the male end of a ¾-in. drive-extension bar can be used to wind the sheave. With the tangs bent down into light rubbing contact, rotate the sheave and wind the spring tight. With your free hand, guide the spring through the slot in the shroud. Press the notched end of the spring into engagement with the anchor slot.

Without releasing the sheave, thread the rope into it. A length of piano wire can be used as a pilot. Fish the end of the rope through the knot hole, tie it, and seal the frayed edges with a match. Push the knot down into the hole for clearance. The process is the same with new-style pulleys except there is no lug to frustrate your work.

Secure the handle with a figure eight knot, leaving about ¾ in. of rope beyond the knot. Seal the end with heat and slip the handle pin through one of the knot loops.

Release the spring in a controlled manner and allow the rope to wind. Bend the lugs so that the nylon bumpers are ¹⁄₁₆ in. below the sheave (the bumpers were against the sheave during winding for better control). Install the shroud assembly on the engine, centering it over the flywheel. Test the starter. If it is slow to retract or binds, loosen the shroud and reposition it.

Vertical pull starter   The vertical pull starter is a convenience on vertical crankshaft engines because it eliminates the need to crouch alongside the engine to start it. This starter is considered a safety feature on rotary lawnmowers. Pulling on the rope sends a nylon gear into engagement with the ring gear on the underside of the flywheel. The nylon gear moves on a thread by virtue of a friction spring and link (an arrangement reminiscent of that used on bicycle coaster brakes). Once the engine fires, the gear retracts back down the thread.

Warning: Wear safety glasses when servicing this and other starters.

The main spring is under some tension. Disarm the spring by lifting the rope out of the sheave groove and winding the sheave, together with the freed section of rope, several turns counterclockwise (Fig. 5-15). When you are finished, there should be no tension on the sheave and approximately 12 inches of rope should be free. Observe the warning stamped on the plastic starter cover and, using a screwdriver, gently pry the cover off. Do not pull on the rope with the cover disengaged.

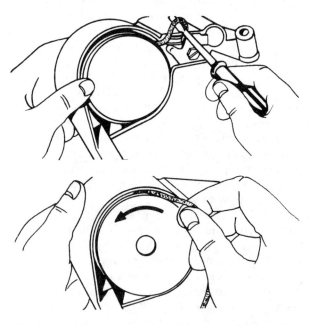

5-15 *Disarm the Briggs vertical pull starter by lifting a foot or so of rope out of the pulley groove.*

Remove the anchor bolt and anchor and note how the spring mates with it (Fig. 5-16). If the spring is to be replaced, carefully work it out of the housing. Remove the rope guide and observe the position of the link (Fig. 5-17) for assembly reference. Using a piece of piano wire in conjunction with long-nosed pliers, pull the rope far enough out of the sheave to cut the knot. Clean mechanical parts in solvent. The friction spring and link are the most vulnerable elements in this mechanism. See that the link and spring assembly move the drive gear to its extremes of travel. If there is any hesitation, replace these parts.

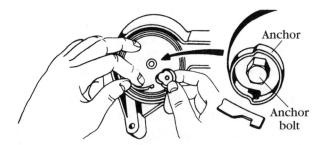

5-16 *Remove the anchor bolt and spring.* Briggs & Stratton Corp.

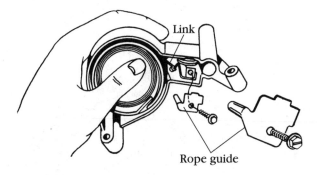

5-17 *Observe the position of the friction link before disassembly.*
Briggs & Stratton Corp.

Begin reassembly by installing the spring in its housing. Slip one end into the retainer slot and wind the spring counterclockwise (Fig. 5-18). Using a length of piano wire or a jeweler's screwdriver, snake one end of the rope into the pulley.

5-18 *To install the spring, anchor it in the retainer slot and wind counterclockwise.* Briggs & Stratton Corp.

Extract the end of the rope from behind the sheave and tie a small, hard knot. Space is critical. No more than ¹⁄₁₆ of an inch of rope should extend beyond the knot. Melt the ends with a flame and wipe down the melted fibers with a shop rag to reduce their diameter. Pull the rope tight and check that the knot clears the threads.

Install the rope guide with the link positioned as it was originally found (Fig. 5-19). Wind the spring counterclockwise with your thumbs to retract the rope (Fig. 5-20). Once the handle butts against the starter case, secure the spring anchor with 80 to 90 lb./in. of torque. Lightly lubricate the spring with motor oil.

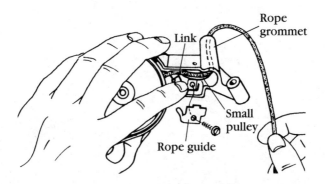

5-19 *Install the friction link behind the rope guide.* Briggs & Stratton Corp.

Snap the starter cover into place and disengage approximately 12 inches of rope from the sheave. Give the rope and sheave two or three clockwise turns to preload the main

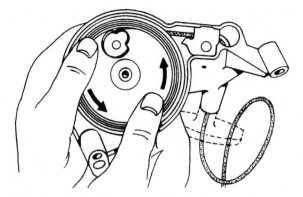

5-20 *Wind the spring counterclockwise to retract the rope.* Briggs & Stratton Corp.

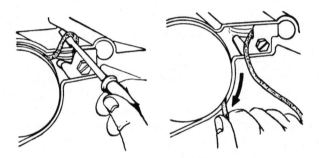

5-21 *Preload the spring two or three rotations.* Briggs & Stratton Corp.

spring and to assure that the rope will be rewound smartly (Fig. 5- 21).

Impulse starters

The spring-powered impulse starter should not be given new life from repairs. These devices were inefficient, generally unreliable, and quite dangerous. In most cases, a B & S side-pull rewind starter can be easily substituted. However, the impulse starter must be at least provisionally disarmed.

Release tension by placing the control knob or remote-control lever in the "start" position. If the engine is locked and the starter fails to unwind, turn the knob or lever to the "crank" position. Get a strong grip on the crank handle with one hand and remove the Phillips screw at the top of the assembly.

Warning: If unsecured, the crank handle can spin with enough force to break an arm.

Dispose of the old starter without dismantling it further or handling the parts more than necessary. The main spring retains a serious potential for bodily damage.

Starter motors—various models

In recent years, Briggs has used a variety of starter motors, manufactured internally and purchased from second parties. Both 12 Vdc and 120 Vac models are supplied. The following discussion covers the more popular models but is not all-inclusive.

Does not crank	low battery; low line voltage (120 Vac); high resistance connection; open starter switch; heavy load; defective motor or rectifier (120 Vac)
Cranks slowly	low battery; low line voltage (120 Vac); high resistance connection; worn motor bearings; worn or sticking brushes; heavy load

Figure 5-22 illustrates three starter motors available for the 140000-, 170000-, and 190000-series engines. These motors are typical of all gear-driven types of motors.

Briggs & Stratton suggests two test parameters, no-load rpm, and no-load current draw, for the motors that dealer mechanics service. To perform these tests, you will need a hand-held tachometer, an ammeter, and a power supply. Depending on the starter motor, the power supply is a fully charged 6 V or 12 V lead-acid battery or 12 V Nicad battery, or a 120 Vac source. The current readings in Table 5-2 are steady draw readings—disregard initial surges.

Mark the end cap and motor frame for assembly reference and remove the two through-bolts that secure the cap to the frame. Take off the brush cover and cap. The armature can be withdrawn from the drive side with the pulley still attached. Starter motor failure can be traced to:

- binding (scored or dry) armature shaft bearings
- worn armature shaft bearings
- shorted, opened, or grounded armature
- shorted, opened, or grounded field
- brushes worn to half or less of their original length
- brushes sticking in their holders.

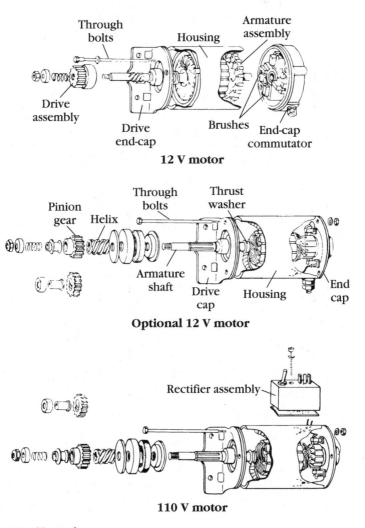

12 V motor

Optional 12 V motor

110 V motor

5-22 *Typical starter motors.* Briggs & Stratton Corp.

Reddish brown discolorations on the commutator bars are
normal and mean that the brushes have seated. Burned commutator bars signal a shorted winding. Glaze and minor imperfections can be removed with number 00 sandpaper as shown
in Fig. 5-23. Severe out-of-round, deep pits, or scores should be
corrected with a lathe. After any of these operations, cut down
the mica with a tool designed for this purpose, or with a nar-

Table 5-2. Steady draw current ratings

Motor	Engine model	Minimum (rpm)	Maximum current draw	Power supply
12 Vdc geared	140000, 170000, 190000	5000	25.0A	6 V battery
12 Vdc geared (American Bosch No. SMH 12A11)	140000, 170000, 190000	4800	16.0A	12 V battery
110 Vac geared	140000, 170000, 190000	5200	3.5A	110 Vac
12 Vdc geared	130000	5600	6.0A	12 V battery
110 Vac geared	130000	8300	1.5A	110 Vac
12 Vac geared	300400, 320400	5500	60.0A	12 V battery
12 Vdc geared (Nicad)	92000, 110900	1000	3.5A	12 V Nicad battery

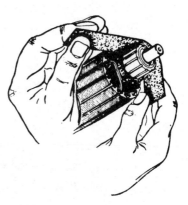

5-23 *Cleaning the commuta-tor.* Tecumseh Products Co.

row, flat-edged jeweler's file (Fig. 5-24). Polish the commutator to remove burrs and clear the filings with compressed air.

Bearings are the next most likely area of failure. The starter might turn freely by hand, but when engaged against the flywheel groan through a revolution or so, and then bind.

Drive out the old bushings, being careful not to score the bearing bosses, and drive in new ones to the depth of the originals. Bushings in motor end covers can be removed by any of several methods. A small chisel can be used to split the bushing. American Bosch end-cover bushings have a flange to accept thrust loads that can be used as a purchase point to collapse the bushing inward. The neatest technique is to pack the boss with heavy grease, then ram the bushing out with a punch that matches the diameter of the motor shaft. A sharp hammer blow will lift the bushing by hydraulic pressure.

The armature can develop shorts. Check for shorts between the shaft and armature with a 120 Vac test lamp. All iron and steel parts must be electrically isolated from nonferrous (brass or copper) parts. Check adjacent commutator bars by the same method. Handle the 120 Vac probes with extreme caution—holding one in each hand means that an electric current could pass through your vulnerable thorax.

Internal winding-to-winding shorts can be detected with a growler. These tools are fairly expensive to buy, but a few auto-parts houses keep one for customer use. You can build one around the core of a television power supply transformer.

If one of the armature windings is shorted, a hacksaw blade will vibrate when placed over the affected armature segment (Fig. 5-25). An open winding will generate sparks between the

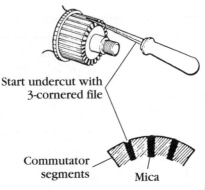

Start undercut with
3-cornered file

Commutator
segments

Mica

5-24 *Undercutting
the mica.* Kohler of Kohler

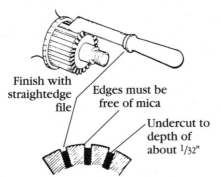

Finish with
straightedge
file

Edges must be
free of mica

Undercut to
depth of
about 1/32"

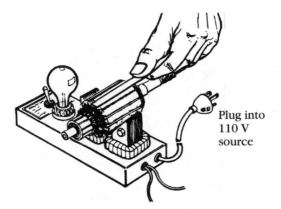

Plug into
110 V
source

5-25 *Checking for internal (winding-to-winding) shorts with a
growler and hacksaw blade.* Tecumseh Products Co.

blade and adjacent commutator segments. Unless the damage is visible, such as a broken connection between the armature and a commutator bar, there is no practical way to repair an armature. Rewinding costs more than a replacement.

American Bosch motors use permanent magnetic fields that require no service under normal circumstances. Arc welding on adjacent parts or extreme vibration can weaken the magnets. Few shops have the necessary equipment to "recharge" magnets and the fields must be replaced.

Brushes must be at least half their original length to maintain pressure against the commutator bars. One brush or brush set should be grounded to the frame, while the remaining brush or brush set connects to the armature. The brushes should be free to move in their holders, and the assembly must be free of carbon dust and oil deposits. In most cases, the brushes must be "shoe-horned" over the commutator with a homemade tool (Fig. 5-26).

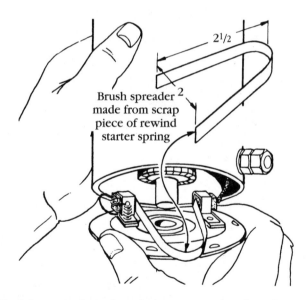

5-26 *A homemade tool used to overcome brush-spring tension during reassembly.* Briggs & Stratton Corp.

6

Charging systems

Engine models employ more than a dozen distinct alternators for use with lead-acid batteries. The Nicad system is recharged from house current through a step-down transformer and rectifier.

Storage batteries

Storage batteries can fail mechanically or electrically. The leading causes of mechanical failure are loose battery hold-down hardware, poor vibration insulation, and owner abuse. The battery straps—the internal busses that connect the cells—are cast as part of the terminals. Twisting the cables or overtightening the terminal bolts can fracture the straps.

Electrical failure is usually associated with chronic low states of charge. The plates become impregnated with sulphate crystals and are no longer capable to take part in the ion exchange that generates electrical potential. A partial cure is to trickle charge the battery for a week or more. Some of the sulphate dissolves into solution. However, the best cure is prevention. Distilled water should be added to cover the plates with electrolyte, and the state of charge should be held greater than 75 percent, or 1.1220 on a temperature-corrected hydrometer.

Deep-charge/discharge cycles encourage sulphation and, if the system is not properly regulated, can overheat the battery and melt holes in the plates. The extent and rate of discharge can be reduced by keeping the battery charged and the engine in tune. The less cranking the better, particularly if the battery shows signs of fatigue. Self-discharge can be controlled by frequent transfusions and by keeping the battery top and terminals clean. The rate of charge is, practically speaking, beyond owner control, although it is wise to invest in an ammeter to keep an eye on the system.

Before turning to specific test procedures, it should be noted that the capacity of the battery has some bearing on its longevity. All things being equal, a larger battery will outlive one that delivers its last erg of energy each time the engine is cranked. But the capacity of the battery, usually measured in ampere hours (A or Amp), cannot compensate for long-term withdrawals. Ultimately, even the large battery must be recharged depending on the output of the alternator and the way the engine is used. A small alternator that is adequate for one start a day might not deliver the current for 20 starts a day.

The first evidence of charging-system failure is a low battery. The state of charge—how much potential is available in the battery—is easily measured with a hydrometer. While hydrometer results do not take the place of a performance test, the hydrometer is the instrument to be tried first.

A hydrometer consists of a squeeze bulb, a float chamber, and a precisely weighted float (Fig. 6-1A). The float is calibrated in units of specific gravity. Water has a specific gravity of 1.000. Sulphuric acid, the other ingredient of electrolyte, has a specific gravity of 1.830. In other words, sulphuric acid is 1.830 times heavier than an equal amount of water. The amount of acid in the electrolyte reflects the state of charge. The more acid in the electrolyte, the greater the charge and the heavier the electrolyte. Each cell in a fully charged battery should have a specific gravity of 1.240–1.280. A completely discharged battery will have a specific gravity of about 1.100.

Draw enough electrolyte into the hydrometer to set the float adrift. The float must not touch the sides of the instrument. Sight across the main level of the instrument. Disregard the meniscus that clings to the sides of the chamber and record the specific gravity for that cell. Repeat the operation on the other cells. The battery should be suspected of malfunctioning if any of the cells fall five points (0.005) below the average of the others.

While raw, uncorrected readings are generally adequate, it should be remembered that acid and water expand when heated. The higher the temperature of the electrolyte, the lower the apparent specific gravity. Expensive hydrometers sometimes incorporate a thermometer in the barrel and a temperature-compensating scale. Any accurate thermometer will work. For each 10 degrees above 80 degrees Fahrenheit add four

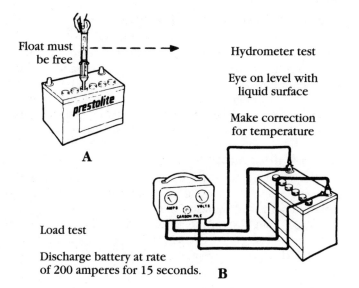

Float must
be free

Hydrometer test

Eye on level with
liquid surface

Make correction
for temperature

A

Load test

Discharge battery at rate
of 200 amperes for 15 seconds. **B**

6-1 *A battery hydrometer and rheostat OMC.*

points (0.004) to the reading. Subtract four points for each 10 degrees less than 80 degrees Fahrenheit.

The most reliable field test requires a carbon pile (Fig. 6-1B) or a rheostat and a voltmeter. The temperature-compensated specific gravity should be at least 1.220 to prevent battery damage. Connect the voltmeter across the terminals and adjust the load to three times the ampere-hour rating. For example, the carbon pile should be adjusted to discharge a 30 A battery at the rate of 90 amps. Continue to discharge for 15 seconds. At no time during the test should the voltmeter register less than 9.6 V. If it does, the battery should be suspected of malfunctioning.

Alternators

Briggs & Stratton provides battery-charging current with an engine-driven alternator and solid-state rectifier. The rectifier converts alternating current into pulsating direct current. The more sophisticated systems include a solid-state voltage regulator to protect the battery from overcharging and to extend headlamp life. Optional features include an ammeter and an isolation diode. All systems employ a lead-acid storage battery.

System 3 & System 4 alternators

Model 90000 and 110000 engines currently use System 3 and 4 alternators, which employ two coils and a full-wave rectifier on a dedicated stator. Versions available are 6 Vdc and 12 Vdc. Output should be 0.5 A at 2800 rpm, as measured between the battery lead and engine ground. Replace the alternator if the output is under specification. Set the stator-to-flywheel air gap for ignition coils at 0.010 in., as described in chapter 3.

The 0.5 A alternator

The 0.5 A alternator is the peewee of the series, consisting of a single charging coil and integral rectifier. Found on model 120000 vertical and horizontal crankshaft engines, the unit should deliver 0.5 A at 2800 rpm. If it fails to produce its rated output, replace the assembly. The early version used a pinch bolt, since discarded, for air gap adjustment, which is set at 0.07 in.

The 1.5 A alternator

The three-coil 1.5 A alternator is used on 130000 series engines (Fig. 6-2) where it is coupled to a 12 A or 24 A battery. Note that there are two styles of connector, both of which contain a soldered-in rectifier, available as a replacement part. Other than

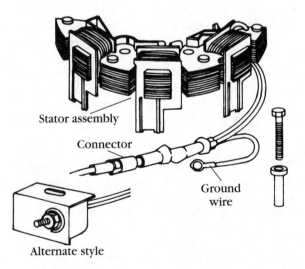

Stator assembly

Connector

Ground wire

Alternate style

6-2 *1.5 A alternator found on 130000 engines. Note the variance in rectifier type.* Briggs & Stratton Corp.

the resistance provided by the battery, the circuit has no voltage or current regulation.

To test the alternator output, connect a number 4001 headlamp between the rectifier output and a paint-free engine ground (Fig. 6-3). The battery must be out of the circuit. Under no circumstances should the output of this or any other alternator be grounded. To do so is to invite burned coils and fried diodes.

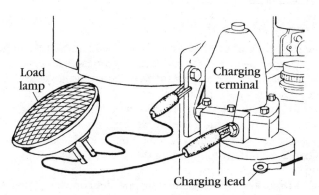

6-3 *Testing output of the 1.5 A alternator.* Briggs & Stratton Corp.

If the lamp refuses to light, the fault is in the rectifier or the alternator. Test the rectifier first because it is the more likely failure point. With the engine stopped, touch the probes of a low-voltage ohmmeter to the output terminal and ground as shown in Fig. 6-4. You should get continuity in one direction and high resistance in the other. If not, replace the rectifier box.

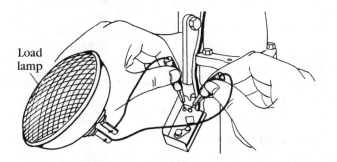

6-4 *Testing stator of the 1.5 A alternator.* Briggs & Stratton Corp.

Test the stator with a 4001 headlamp connected across the output leads (Fig. 6-4). The lamp should burn. If not, check the leads to the stator for possible fouling. Before deciding that the stator is defective, compare the magnetic strength of the flywheel ring against one known to be good. Failure is exceedingly rare but not impossible.

Install a replacement stator and torque the cap screws 18 to 24 lb./in. See that the output leads are snug against the block and well clear of the flywheel.

dc-only & ac-only alternators

The dc-only alternator is found on 170000 and larger engines; the ac version on 190000 and larger engines. An identical four-coil stator produces 5 A for the ac version and, because of rectifier resistance, 4 A for the dc unit (Fig. 6-5). The ac alternator should produce 14 Vac no-load at 3600 rpm. If not, replace the stator. Current output from the dc version, as measured by an ammeter in series with the charging lead, should range between 2 A and 4 A, depending on battery voltage. If no or low output is indicated, test the in-line diode with an ohmmeter. It should register low resistance in one direction and high resistance in the other. If the diode tests okay, the stator is at fault.

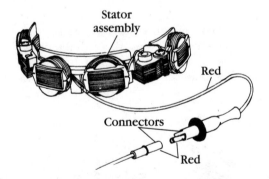

6-5 *The dc-only rectifier can be distinguished from its ac-only twin by the bulge in the connector, signaling the presence of a diode rectifier.* Briggs & Stratton Corp.

The 4 A alternator

Used on 17 and 190000 engines, the 4 A alternator has eight charging coils arrayed on a 360 degree stator. This alternator does not include a regulator, but is otherwise identical to the 7 A type illustrated in Fig. 6-5.

Troubleshooting procedures begin with a short-circuit test. Connect a 12 V test lamp between the rectifier output and the positive terminal of a charged battery (Fig. 6-6). If the lamp lights, battery current is being fed back to ground through the charging circuit. Unplug the rectifier connection under the blower housing. If the lamp goes out, the rectifier is okay and the problem lies in the alternator and associated wiring. If the lamp continues to burn, the rectifier is at fault and must be replaced.

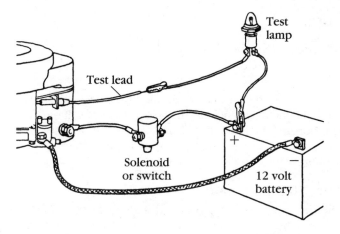

6-6 *Testing for shorts in the 4 A, 7 A, and dual-circuit alternators.* Briggs & Stratton Corp.

Inspect the output leads from the alternator for frayed insulation or other evidence of shorts before you replace the stator assembly. Make necessary repairs with electrical tape and silicone cement.

This alternator has four distinct windings, each involving two coils. A break in one of the windings drops output by a third. Check each of the four pins with the fuse-holder lead as shown in Fig. 6-7. Each pair of pins supplies current to a diode in the rectifier. Should a diode blow, a quarter of the output is lost. Check each of the four rectifier terminals with an ohmmeter. One probe should be on a good (paint-free and rust-free) ground on the underside of the blower housing; the other

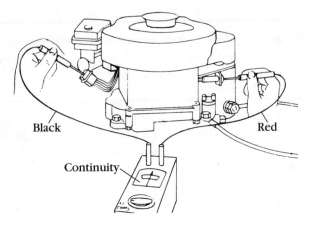

6-7 *Testing the stator in the 4 A and 7 A alternator.* Briggs & Stratton Corp.

probe should be on one of the diode connection points. Observe the meter and reverse the test connections. If the diode in question is functional, it will have continuity in one direction and very high resistance in the other. Repeat the test for the remaining three connection points.

The 7.0 A alternator

The 7.0 A alternator is used on series 140000, 170000, and 190000 engines and can be easily recognized by the connector plug, which is flanked by a regulator on one side and the rectifier on the other (Fig. 6-8). Some installations employ an isolation diode in a tubular jacket on the outside of the shroud. The purpose is to block current leakage from the battery to ground by way of the alternator windings. Applications that do not have this diode isolate the battery at the ignition switch.

Test the isolation diode by connecting a 12 V lamp in series with the output (Fig. 6-9). The lamp should not light. If it does, the diode is shorted and must be replaced. Check diode continuity with an ohmmeter connected between the two diode leads. The meter should show zero resistance in one direction and high resistance when the leads are reversed.

To test the stator, regulator, or rectifier, connect a test lamp as shown previously in Fig. 6-6. Do not start the engine. If the lamp lights, one of the three is shorted. Disconnect the rectifier-regulator plug under the blower housing and remove the stator

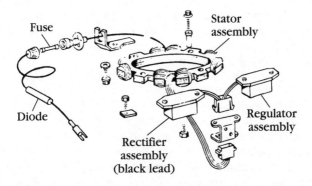

6-8 *The 7.0 A alternator configuration.* Briggs & Stratton Corp.

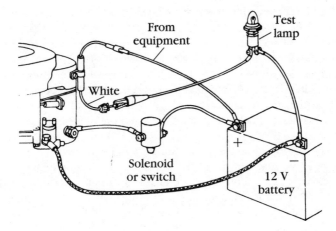

6-9 *Testing the isolation diode on the 7.0 A alternator.* Briggs & Stratton Corp.

from the circuit. If the lamp continues to burn, the regulator or rectifier is shorted. Test these two components individually to determine which is at fault.

Test the rectifier as described previously. Two black leads, joined by a connector, go to the rectifier. Each lead services two pins on the rectifier side of the connector. Without removing the rectifier assembly from the shroud, connect ohmmeter leads between each of the four pins and a paint free ground on the underside of the shroud. Observe the meter reading at each pin and reverse the leads. The pins should conduct in one direction and not in the other. If current flows in both directions, the rectifier is shorted. If no current passes, the rectifier is open. In either event, the assembly must be replaced. Instructions are packaged with the new part.

The regulator is distinguished by one red and one white lead. Test as above. The white lead pin must show some conductivity in one direction and none, or almost none, in the other. The red lead pin should give no reading in either direction. If it is necessary to replace the regulator, instructions are supplied with the replacement part.

Check stator continuity as shown previously in Fig. 6-7. Each of the four pins must be contiguous with the lead at the fuse holder. If not, check the visible wiring for defects before you invest in a new stator.

The dual-circuit alternator The dual-circuit alternator is one of the most interesting alternators used on Briggs & Stratton engines. Two stator windings are provided, one for the headlights and the other for the battery. Battery output is rectified and rated at 3 A. Headlight output is alternating and can deliver 5.8 A at 12 V at wide-open throttle. Two versions of this alternator, one with a single plug and the other with separate ac and dc output lines, are used on 170000 and Lorgen engines.

The battery circuit is protected by a 7.5 A type AGC or 3AG automotive fuse and might be supplied with an ammeter. The ac circuit is independent of the charging circuit, although good practice demands that both be grounded at the same engine mounting bolt. Each circuit is treated separately here.

Charging circuit Check output with an ammeter in series with the positive battery terminal (Fig. 6-10). The meter should show some output at medium and high engine speeds. No charge indicates a blown fuse, shorted or open wiring, or a failed rectifier or alternator.

Connect a 12 V test lamp between the battery and the charging section as illustrated in Fig. 6-6. The lamp should not light. If it does, the alternator or rectifier is defective. To isolate the problem, disconnect the plug under the shroud. If the light goes out, the rectifier is good and the difficulty is in the alternator or its external circuitry. If the light continues to burn, the rectifier is shorted to ground and must be replaced.

To test the stator, remove the starter motor, shroud, and flywheel. Inspect the red output lead for frayed or broken insu-

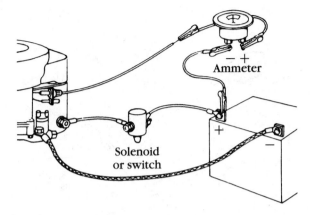

6-10 *Testing dc output on the dual-circuit alternator.* Briggs & Stratton Corp.

lation. Repair with electrical tape and silicone, being careful to route the lead away from moving parts. Test the stator by connecting an ohmmeter between the terminal at the fuse holder and the red lead pin in the connector. The meter should show continuity. If not, the stator is open and must be replaced.

Test for shorts by connecting the ohmmeter between a good ground and each of the three black lead pins in sequence (Fig. 6-11). Test for continuity by holding the probes against the two black pins (Fig. 6-12). If the circuit is open, the stator is good. If the meter shows continuity, the stator must be replaced.

The rectifier mounts under the fan shroud where it is serviced by a three-prong connector plug. Open the plug and connect one test lead from an ohmmeter to the red lead pin and the other to the underside of the shroud. Observe the meter and reverse the test leads. The meter should report high resistance in one direction and no resistance in the other. Do the same for each black lead pin.

The lighting circuit should be tested with a number 4001 headlamp connected between the output terminal and a reliable engine ground. The lamp should burn brightly at medium engine speeds. If it does burn brightly, the problem is in the external circuit between the engine and the vehicle lights. If the lamp does not light or burns feebly, the problem is in the alternator. Check coil continuity with an ohmmeter as shown in Fig. 6-13. High or infinite resistance means a defective stator.

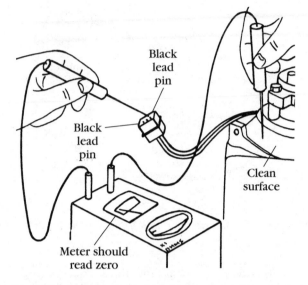

6-11 *Testing for a shorted charging coil on the dual-circuit alternator.* Briggs & Stratton Corp.

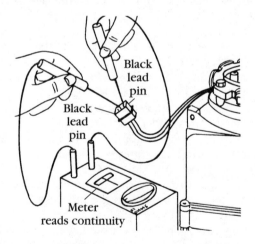

6-12 *Testing charging-coil continuity on the dual-circuit alternator.* Briggs & Stratton Corp.

The 10 A alternator

Used on the series 200400 and 320400 engines, the 10 A alternator is a heavy-duty device delivering better than 4 A at 2000

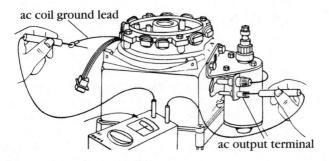

ac coil ground lead

ac output terminal

6-13 *Testing ac circuit continuity on the dual-circuit alternator.* Briggs & Stratton Corp.

rpm and a full rating at 3600 rpm. The regulator is more flexible than those used on the smaller engines and can handle large loads without overcharging the battery.

Check voltage across the battery terminals with the engine turning at full-governed rpm. Less than 14 V on a fully charged battery means stator or regulator rectifier problems.

Disconnect the plug at the regulator rectifier and connect an ac voltmeter to each of the two outside plug terminals (Fig. 6-14). A reading of less than 20 V per terminal means a defective stator. Check the regulator rectifier by default. That is, if the system fails to deliver sufficient charging voltage and the stator appears okay, replace the regulator rectifier.

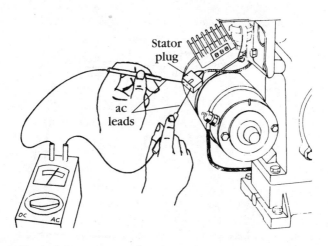

Stator
plug

ac
leads

6-14 *Testing the stator on the alternator.* Briggs & Stratton Corp.

The Nicad system

An option on 92000 and 110900 engines, the Nicad system consists of a gear-driven starter motor, starter-ignition switch, plug-in battery charger, and a 12 V nickel-cadmium battery. Nicad systems are intended for rotary lawnmower applications so the starter-ignition switch is mounted on the handlebar where it is electrically isolated from the engine. The switch stops the engine by grounding the magneto primary circuit through the connector clipped on the engine shroud. If the connector comes free of its clip, the magneto will be denied ground but the engine will continue to run regardless of the switch setting.

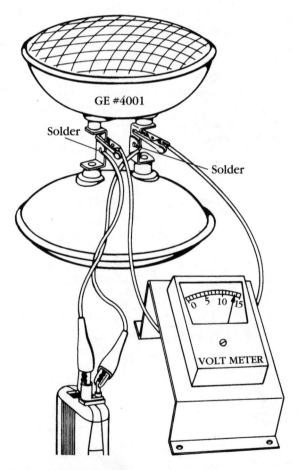

6-15 *Testing the Nicad battery.* Briggs & Stratton Corp.

The first place to check if trouble arises on these engines is the battery. Nickel-cadmium cells are by no means immortal. Load the battery with two G.E. number 4001 sealed-beam head-lamps connected in parallel (Fig. 6-15). Monitor the voltage. The meter should show at least 13.6 V after one minute of draw. Readings of 13 V and less mean that one or more of the cells are defective. The lights should burn brightly for at least five minutes.

The half-wave rectifier supplied with this system should be capable of recharging a fully depleted battery over a period of 16 hours. An inexpensive tester can be constructed from the following materials:

- 1 1N4005 diode
- 1 red lamp socket (Dialco No. 0931-102)
- 1 green lamp socket (Dialco No. 0932-102)
- 1 neon bulb, No. 53
- 1 ¾-in. machine screw, No. 6-32
- 1 ¾-in. machine screw, No. 3-48

Wire the components as shown in Fig. 6-16. If neither bulb lights, the transformer or charger diode is open. If both bulbs light, the charger diode is open and passing alternating current. A properly working charger will light only the green bulb.

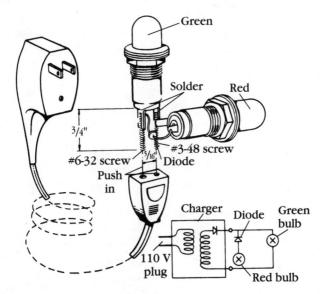

6-16 *A homemade rectifier tester.* Briggs & Stratton Corp.

7

Engine mechanics

This chapter describes repair and overhaul procedures for side-valve, four-cycle, and 95/96700-series two-cycle engines. In addition to mechanic's handtools (U.S. inch standard), you should have access to:

- 0–100 lb./ft. torque wrench
- 0–250 lb./in. torque wrench
- 0–6 in. machinist's calipers.

Most special tools called for in the text can be fabricated from steel plate and pipe or purchased from a factory distributor. You will also need a solvent degreaser, such as Varsol (which costs only slightly more than kerosene and works better), a supply of lintless rags or heavy-duty paper towels, and a clean, flat work area large enough to lay out parts in the order of removal.

Vertical-crankshaft engines are awkward to handle while the crankshaft remains in place. The holding fixture shown in Fig. 7-1 provides a stable platform clear of bench clutter.

Diagnosis

Begin by trying to assess the amount of damage. Major areas of concern are:

- *Black, carbonized crankcase oil:* expect to find severe and possibly noncorrectable wear on all bearing surfaces.
- *Leaking air-cleaner gasket or punctured cleaner element:* expect to hone the cylinder on all engines; two-cycle bearings will also be contaminated and might exhibit severe wear.

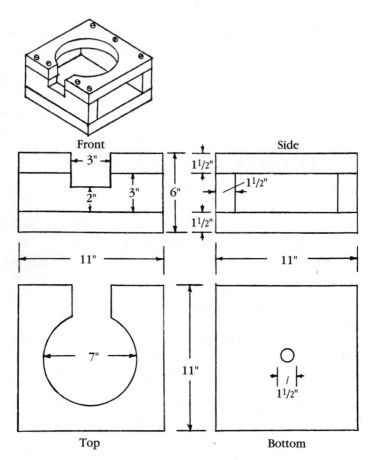

7-1 *A work stand for vertical shaft engines can be constructed easily from scrap lumber.*

- *Loss of power:* in two-cycle engines, loss of power is nearly always associated with gas leaks past the piston rings. See Fig. 7-2 for four-cycle engines.
- *High oil consumption:* see Fig. 7-3 for a list of probable causes.
- *Bent crankshaft:* visible wobble means the crankshaft must be replaced.

Note: The factory position is that bent crankshafts should not be straightened because of the risk of subsequent failure. If the crankshaft in question is attached to a lawnmower blade, the results could be tragic.

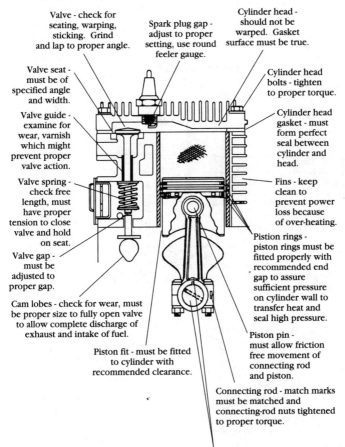

Ignition - must be properly timed so that spark plug fires at precise moment for full power.

Valve - check for seating, warping, sticking. Grind and lap to proper angle.

Spark plug gap - adjust to proper setting, use round feeler gauge.

Cylinder head - should not be warped. Gasket surface must be true.

Valve seat - must be of specified angle and width.

Cylinder head bolts - tighten to proper torque.

Valve guide - examine for wear, varnish which might prevent proper valve action.

Cylinder head gasket - must form perfect seal between cylinder and head.

Valve spring - check free length, must have proper tension to close valve and hold on seat.

Fins - keep clean to prevent power loss because of over-heating.

Valve gap - must be adjusted to proper gap.

Cam lobes - check for wear, must be proper size to fully open valve to allow complete discharge of exhaust and intake of fuel.

Pistion rings - piston rings must be fitted properly with recommended end gap to assure sufficient pressure on cylinder wall to transfer heat and seal high pressure.

Piston fit - must be fitted to cylinder with recommended clearance.

Piston pin - must allow friction free movement of connecting rod and piston.

Connecting rod - match marks must be matched and connecting-rod nuts tightened to proper torque.

Oil passages - all oil holes and passages must be clear to allow full lubrication for friction-free operation.

Air filter - should be clean to allow engine to breathe.

Carburetor - must be set properly to assure proper and sufficient air and fuel.

7-2 *Factors that affect four-cycle engine power output.* Tecumseh Products.

Cylinder head

Note the bolt lengths. Aluminum block engines typically have longer bolts in the exhaust-valve area (Fig. 7-4). The head usu-

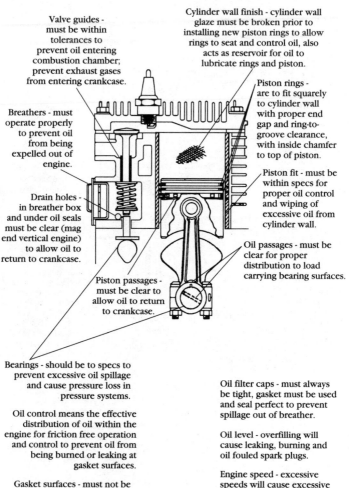

Valve guides - must be within tolerances to prevent oil entering combustion chamber; prevent exhaust gases from entering crankcase.

Cylinder wall finish - cylinder wall glaze must be broken prior to installing new piston rings to allow rings to seat and control oil, also acts as reservoir for oil to lubricate rings and piston.

Breathers - must operate properly to prevent oil from being expelled out of engine.

Piston rings - are to fit squarely to cylinder wall with proper end gap and ring-to-groove clearance, with inside chamfer to top of piston.

Drain holes - in breather box and under oil seals must be clear (mag end vertical engine) to allow oil to return to crankcase.

Piston fit - must be within specs for proper oil control and wiping of excessive oil from cylinder wall.

Oil passages - must be clear for proper distribution to load carrying bearing surfaces.

Piston passages - must be clear to allow oil to return to crankcase.

Bearings - should be to specs to prevent excessive oil spillage and cause pressure loss in pressure systems.

Oil control means the effective distribution of oil within the engine for friction free operation and control to prevent oil from being burned or leaking at gasket surfaces.

Gasket surfaces - must not be nicked, old gasket removed, always use new gaskets.

Oil filter caps - must always be tight, gasket must be used and seal perfect to prevent spillage out of breather.

Oil level - overfilling will cause leaking, burning and oil fouled spark plugs.

Engine speed - excessive speeds will cause excessive oil consumption by burning and leaking.

7-3 *Factors that affect four-cycle engine oil consumption.* Tecumseh Products Co.

ally drops off the block once the bolts have been removed. If it sticks, break the gasket seal with a rubber mallet. Inspect the head gasket for breaks and gasket mating surfaces for carbon tracks. A blown head gasket is usually an isolated phenomenon, but it can mean a warped head or pulled bolt threads. Spark plug boss and head bolt threads can be repaired with Heli-Coil or factory-supplied kits:

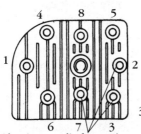

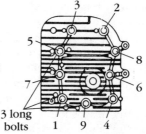

Aluminum cylinder engines
(15 cu. in. and less
except 100700) long screws
in these 3 holes

Aluminum cylinder engines
(17, 19, 22, & 28 cu. in.)

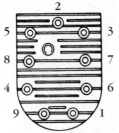

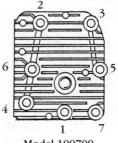

Models 23 - 230000
240000 - 300000 - 320000
iron

Model 100700

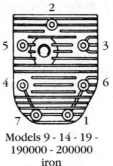

Models 9 - 14 - 19 -
190000 - 200000
iron

7-4 *Torque sequences for side valve Briggs & Stratton engines.*

| M14 × 1.25 spark-plug thread repair kit | PN 100013 |
| $\frac{5}{16}$–$\frac{5}{18}$ UNC head bolt kit | PN 100016 |

Remove the carbon deposits, if the engine is not going to be further disassembled, from the piston top. Scrape off any gasket fragments that remain on the mating surfaces.

Warning: Engines built as recently as the early 1970s used asbestos head gaskets. Remove gasket material with a single-edge razor blade and dispose of the shards safely. Do not use a wire wheel or any other method that might create dust.

It is good practice to check the head for distortion (Fig. 7-5). Using a piece of plate or mirror glass as a work table, try to insert a 0.003-in. feeler gauge between bolt holes. If the gauge passes, the head should be resurfaced. An "Armstrong mill" is the least expensive way to do this. An Armstrong mill can be made by taping a sheet of #360 wet-or-dry abrasive to the glass.

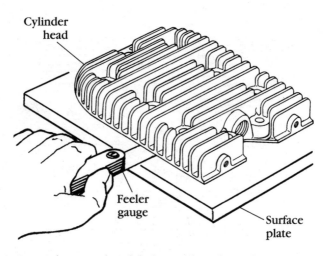

7-5 *Overtightening head bolts tend to bow the casting and invite leaks. Resurface if a 0.002-in feeler blade can be inserted past the gasket flange and surface plate.* Kohler of Kohler

Clean bolt threads and lubricate with graphite grease or assembly lube before installation. Figure 7-4 illustrates torque sequences for the various models. Run down the bolts in three increments: one-third, two-thirds, and full torque. Torque limits are listed in Tables 7-1 and 7-2.

Flange

The flange, or the power-takeoff-side (PTO) crankcase cover, seals the crankcase, supports the PTO end of the crankshaft, and on most models, locates the oil slinger. The flange might also incorporate a camshaft-driven PTO. Remove the sheave or

Table 7-1.
Aluminum block, side-valve,
single-cylinder engine torque limits

Displacement (cubic inch)

6	8	9	10	11	13	14	17	19	22/25	28

Flywheel nut (lb/ft)

55	55	55	60*	55	60	65	65	65	65	65

Cylinder head (lb/in)

140	140	140	140	140	140	165	165	165	165	165

Connecting rod (lb/in)

100	100	100	100	100	100	165	165	185	185	185

Crankcase cover or flange (lb/in)

85	85	85	**	85	120	140	140	140	140	140

 * 100700 flywheel nut 55 lb/ft
** 85 lb/in for 100700 and 120 lb/in for 100200 & 100900

Table 7-2.
Iron block
single-cylinder engine torque limits

Displacement (cubic inches)	23	24	30	32
Flywheel nut (lb/ft)		145		
Cylinder head (lb/in)		190		
Connecting rod (lb/in)		190		

blade adaptor, together with locating keys and other bits of hardware that are secured to the crankshaft-PTO stub. Blade adaptors can be quite stubborn and removal might require the use of a gear puller and liberal quantities of penetrating oil. A long bolt inserted into the center bolt hole protects crankshaft threads from puller damage. Occasionally, an adaptor will defeat even the strongest puller, and it will be necessary to drive the culprit off with simultaneous hammer blows directed at the adaptor neck, adjacent to the crankshaft.

Using coarse strip paper, remove all traces of rust from the crankshaft stub (Fig. 7-6). Dress keyways with a file because lo-

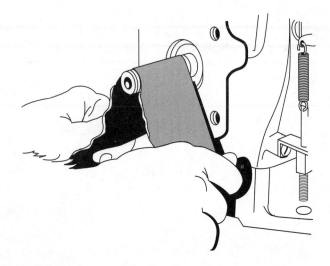

7-6 *Polishing the crankshaft stub.* Clinton Engines Corp.

cal distortion might be present. In addition, the extreme end of
the crankshaft might be expanded enough to score the flange-
side main bearing during disassembly.

Remove flange holddown bolts and lightly oil the crankshaft
stub. Gently separate the flange and crankcase with a mallet.
Prying the flange off will almost certainly damage the gasket
surface with consequent air and oil leaks. Once the flange has
moved enough to disengage the locating pins, it should slip
easily off the crankshaft stub. But do not force matters; if the
flange binds, stop and polish out the high spot.

Thoroughly clean the flange and inspect the main bearing,
which is particularly vulnerable to scoring in vertical crankshaft applications. Ball bearings should turn smoothly and
quietly; plain bearings should be dead smooth, with no more
than local scuffing. It is a peculiarity of aluminum bearings
that the bearing might be ruined without seriously affecting
the crankshaft.

Before installing the flange, align crank and camshaft timing
marks, assemble the oil slinger (with spring on models that require it) and mount a new flange gasket on the block.

The thickness of the flange gasket controls crankshaft endplay on all engine models except those with double ball-bearing mains. Endplay, or the distance the crankshaft can move on
its long axis, is determined after a trial assembly with a new

flange gasket installed. Either a dial indicator or a sheave and feeler gauge is required (Fig. 7-7). Bump the crankshaft to its travel limit in one direction, zero the dial indicator or add leaves to the feeler gauge for zero clearance, and bump the crankshaft as far as it will go in the other direction. Endplay specification for all engines is 0.002 to 0.008 in., except Models 100700 and 120000, which have an allowable range of 0.002 to 0.030 in.

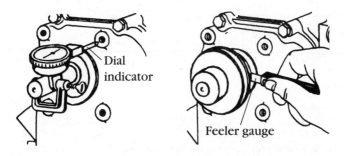

7-7 *Crankshaft endplay is 0.002 to 0.008 in. for all Briggs & Stratton engine models and should be checked upon assembly with a dial indicator or feeler gauge.*

Flange gaskets are available in three thicknesses—0.005 in., 0.009 in., and 0.015 in.—for all engine models. Aluminum block engines require at least one 0.015-in. gasket to prevent oil and air leaks. Cast-iron engines can, when necessary, be assembled with only one 0.005-in. gasket. Adding gaskets increases endplay. However, it is sometimes necessary to reduce endplay, particularly when a new flange is fitted. This can be done by installing a thrust washer between the sump and crankshaft on plain-bearing engines (Fig. 7-8). On aluminum-block engines with a single ball bearing, the thrust washer should be installed at the magneto end of the crankshaft. Engines with two ball-bearing mains will not accept a thrust washer.

Once the crankshaft has acceptable float, remove the flange and lay it on the bench with the oil seal up. Insert a large screwdriver blade under the metal oil-seal lip and pry up (Fig. 7-9). Repeat the operation at several places around the circumference of the seal. A sharp downward blow with the palm of the hand on the screwdriver handle might be required to break the seal free.

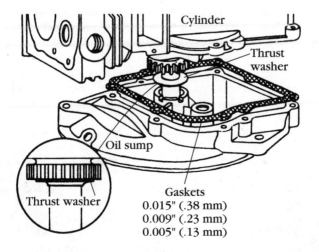

Cylinder

Thrust washer

Oil sump

Thrust washer

Gaskets
0.015" (.38 mm)
0.009" (.23 mm)
0.005" (.13 mm)

220624 Thrust washer for 0.875" (22.23 mm) DIA. CRKSFT.
220708 Thrust washer for 1.0"   (25.40 mm) DIA. CRKSFT.
222949 Thrust washer for 1.181" (30.00 mm) DIA. CRKSFT.
222951 Thrust washer for 1.378" (35.00 mm) DIA. CRKSFT.

7-8 *When a thinner-than-recommended flange gasket won't suffice, a thrust washer is used on plain bearing and single ball-bearing engines to reduce crankshaft endplay. Endplay is increased by either using a thicker-than-normal flange gasket or by stacking flange gaskets.* Briggs & Stratton Corp.

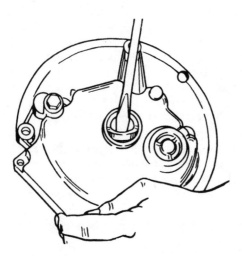

7-9 *Flange oil seal is pried free with the aid of a large screw-driver.* Tecumseh Products Co.

Install a new seal with the steel lip on the outboard side of the flange. Briggs & Stratton supplies dealers with seal installation tools, although a soft wooden block, large enough to cover the whole diameter of the seal, works about as well on plain-bearing engines. Models 60000, 80000, 100000, and 130000 with a ball bearing flange have the seal countersunk ³⁄₁₆-in. below flange depth and require an installation tool sized to match the seal's outside diameter (OD). Grease the elastomer seal lips to prevent damage during installation and startup.

Wipe excess oil from the crankshaft stub and cover the keyways with a single piece of cellophane tape. Keyway edges are sharp and, unless covered, might damage the new seal. Install the correct gasket and flange. Tighten flange holddown bolts evenly, pulling down the casting without distortion. See Tables 7-1 and 7-2 for torque limits.

Oil slingers

Slingers, used as an aid to oil circulation in aluminum block engines, have been manufactured in two basic styles. Early-production slingers mount on a light metal bracket. Replace the slinger if teeth are worn, and bracket and bushing if bracket is worn to 0.49-in. diameter or less. New-style slingers are pivoted on the camshaft, and are replaced as a complete assembly in event of tooth or other damage. Engine Models 100900 and 130900 employ a spring washer on the outboard side of the slinger bracket (Fig. 7-10).

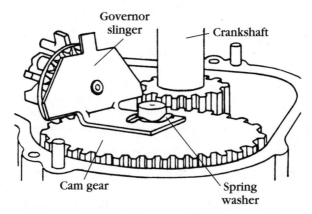

7-10 *A spring washer is used on model 100900 and 130900 oil slingers.* Briggs & Stratton Corp.

Camshaft

This section appears more complicated than the work actually is. Briggs & Stratton employs more than a dozen camshaft variants, each with a more or less unique disassembly and assembly procedure. But the average reader with the average (late-production aluminum-block, plain-bearing, 6- to 9-cubic in.) engine need only remember to align the timing marks before the camshaft is withdrawn and assembled. The rest of this material is for those involved with larger and, for the most part, older engines. Wear limits for all models are shown in Tables 7-3 and 7-4.

Table 7-3.
Cast-iron block, side-valve, and single cylinder engine dimensions (all dimensions in inches)

Displacement				
	23	24	30	32
Valve clearance:				
Intake	0.008	0.008	0.008	0.008
Exhaust	0.018	0.018	0.018	0.018
Crankshaft reject size:				
Mag. journal	1.3769		Ball bearing	
PTO journal	1.3769		Ball bearing	
Crankpin	1.1844	1.3094	1.3094	1.3094
Connecting rod reject size:				
	1.189	1.314	1.314	1.314
Camshaft reject size:				
Journal bearing	0.497	0.497	0.8105/.6146	0.8105/.6146
Lobe	1.184	1.184	1.184	1.184
New bore diameter (± 0.0005)				
	2.9995	3.0620	3.5370	3.5620

Aluminum-block, plain-bearing

To disassemble, remove the flange, align timing marks, and lift out the camshaft. To assemble, insert tappets or valve lifters, align timing marks, and insert the camshaft (Fig. 7-11). Check timing mark alignment. Many crankshafts have removable gears. Install with timing mark visible.

Table 7-4. Aluminum block, side-valve, and single-cylinder engine dimensions (all dimensions in inches)

Displacement (cubic inch)	6	8	9	10	11	13	14	17	19	22/25	28
Valve clearance:											
Intake	0.006	0.006	0.006	0.008	0.006	0.006	0.006	0.006	0.006	0.006	0.006
Exhaust	0.008	0.008	0.008	0.008	0.008	0.010	0.010	0.010	0.010	0.010	0.010
Crankshaft-reject size:											
Mag. journal	0.873	0.873	0.873	0.873	0.873	0.873	0.997	0.997*	0.997*	1.376	1.376
PTO journal	0.873	0.873	0.873	0.998/1.060	0.873	0.998	1.179	1.179	1.179	1.376	1.376
Crankpin (nominal pin journal diameter varies with build date for some models)											
	0.870	0.996	0.997	0.997	0.997	0.997	1.090	1.090	1.122	1.247	1.247
Connecting rod reject size (big-end diameter varies with nominal pin journal diameter):											
	0.875	1.001	1.001	1.001	1.001	1.001	1.095	1.095	1.127	1.252	1.252
Camshaft reject size (iron cams only):											
Journal bearing	0.498	0.498	0.498	0.498	0.436/0.498	0.498	0.498	0.498	0.498	0.498	0.498
Lobe	0.883	0.883	0.883	0.950	0.870	0.950	0.977	0.977	1.184	1.184	
New bore diameter (± 0.0005)	2.3745	2.3745	2.5620	2.5620	2.4995	2.5620	2.7495	2.9995	2.9995	3.4370	3.4370

* 1.179 with synchro-balance

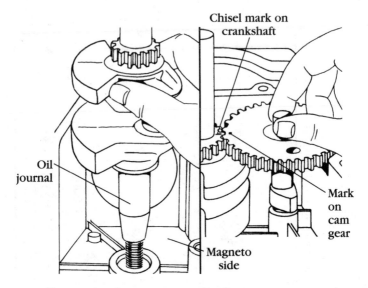

Chisel mark on
crankshaft

Oil
journal

Mark
on
cam
gear

Magneto
side

7-11 *Timing marks are stamped adjacent to gear teeth on plain-bearing engines.* Briggs & Stratton Corp.

Aluminum-block, ball-bearing

Remove the connecting-rod cap (as described in the following section) and align timing marks. Because of the PTO-side ball bearing, the crankshaft timing mark is on the counterweight and not on the timing gear (Fig. 7-12). Remove crankshaft and camshaft as an assembly. To assemble, install well-lubricated tappets, then insert crank and camshaft as an assembly with timing marks indexed.

Cast-iron block, plain-bearing

Remove side plate and connecting rod (detailed in the next section) and turn the crankshaft so that crankpin and counterweight clear the camshaft gear. Withdraw the crankshaft, twisting it as necessary.

The camshaft is hollow and supported by a shaft that extends through both sides of the iron block. Working from the PTO side, use a hammer and blunted ⅜-in.-diameter punch to drive the support shaft out through the flywheel side of the block (Fig. 7-13). Some resistance will be felt as the flywheel-side expansion plug is displaced by the support shaft.

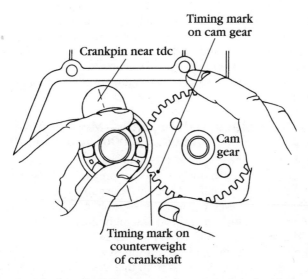

7-12 *Camshaft on ball-bearing engines is indexed with a mark on the crankshaft counterweight.* Briggs & Stratton Corp.

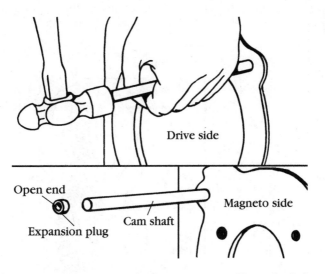

7-13 *Cast-iron engine camshafts are supported by a shaft that is driven out through the magneto side of the crankcase. Note that the expansion plug is installed with its cupped side facing out.* Briggs & Stratton Corp.

Camshaft assembly is essentially the reverse of disassembly. Install the tappets, then insert the loose camshaft and, working from the flywheel side of the engine, install the support shaft. Coat the expansion plug with silicone sealant and hammer home. Install the crankshaft with crank and cam timing marks in alignment.

Cast-iron block, ball-bearing—Models 9, 14, 19, 23, 190000, 200000, 230000, and 240000

These engines are built on the same basic format as the cast-iron, plain-bearing models discussed previously. The camshaft rides on a support shaft, sealed by an expansion plug at the flywheel side of the crankcase. Disassembly procedure is similar, except that the piston and rod assembly must be extracted before the crank is withdrawn on Model 240000.

Because the cam gears are on the flywheel side of these engines, the timing drill varies from that described earlier. Install tappets and camshaft, and with the camshaft nestled in the crankcase recess provided, install the crankshaft (Fig. 7-14). Now align timing marks and, holding the camshaft gear in engagement with the crankshaft gear, install the support shaft from the flywheel side. Lightly tap the support shaft home. Seal

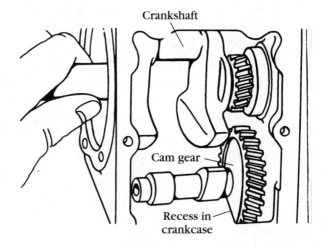

Crankshaft

Cam gear

Recess in crankcase

7-14 *A crankcase recess is provided on cast-iron, ball-bearing engine models to facilitate crankshaft installation. Timing marks are aligned, the camshaft gear meshed with the crankshaft gear, and the support shaft is installed.* Briggs & Stratton Corp.

the expansion plug with silicone and hammer the plug into the flywheel-side support-shaft counterbore.

Cast-iron-block, ball-bearing— Models 300000 and 32000

Remove the short bolt and Bellville washer that secure this gear (Fig. 7-15). Working from the magneto side of the block, loosen the long bolt and Bellville washer two turns. Tap the head of the bolt to free the camshaft support shaft. Turn the bolt out when pushing out the support shaft. Remove two bolts from the cam-gear bearing and, holding the camshaft so that it doesn't drop, remove the camshaft.

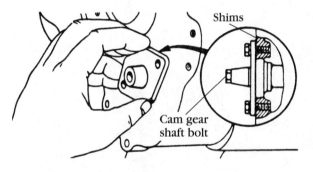

7-15 *Camshaft endplay is fixed by shims on engine models 300400 and 320400.* Briggs & Stratton Corp.

Assembly is quite straightforward. Install tappets and ignition-breaker plunger in their respective bores. Insert camshaft from the PTO side of the block and slide the support shaft (with rounded end toward outside of engine) through the PTO bearing and into the camshaft. Mount the magneto-side bearing and torque-bearing hold-down screws to 85 lb./in. Install 5½-in. support-shaft bolt and tighten. Camshaft endplay specification is 0.002–0.008 in. If endplay is less than 0.002 in. install the appropriate shim:

Shim thickness (in.)	Part number
0.005	270518
0.007	270517
0.009	270516

If endplay exceeds 0.008 in., replace existing bearing with part No. 299706, using shims provided in the kit to establish correct clearance.

Once the crankshaft is installed, it is impossible to see the timing marks on these 30- and 32-cubic in. engines. It is necessary to color the crankshaft gear tooth whose inboard edge is adjacent to the timing mark with crayon or Prussian blue as shown in Fig. 7-16. Install the crankshaft, aligning the referenced tooth with the camshaft timing mark. Main-bearing assembly hold-down screws are torqued to these specifications:

Magneto side	75–90 lb./in.
PTO side	185 lb./in.

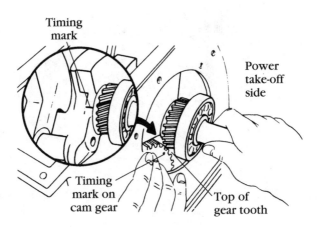

7-16 *Timing marks are not visible after crankshaft installation, and it is necessary to identify the pertinent crankshaft gear tooth with Prussian blue.* Briggs & Stratton Corp.

Connecting rod

Before rod and piston are removed from the engine, it is necessary to carefully scrape off accumulated carbon between the upper limit of piston travel and the top of the piston bore. Chrome-plated bores can be damaged easily. Use a plastic or wooden scraper for this operation. Iron bores might exhibit a pronounced ridge in this area that, if the piston is to be reused, should be removed with a small-engine ridge reamer. J. C. Whitney & Co., the Chicago-based mail-order house, catalogs this tool as part No. 14XX2557A. The cost is about $25.

Rotate the crankshaft until the engine is in time. Note position of the timing marks for future reference. Defeat the connecting-rod lock by using a small punch to straighten the lock tabs enough for wrench purchase (Fig. 7-17). Note the location of any reference marks, oil holes, or other distinguishing characteristics on the big, or lower, end of the rod assembly. Loosen the two connecting-rod bolts in several increments, and rotate the crankshaft a few degrees to disengage the rod cap. If the cap remains in place, tap it lightly on one side with the punch, carefully avoiding damage to parting surfaces.

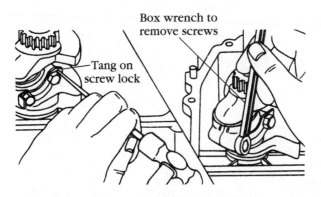

7-17 *Open connecting-rod bolt locks with a punch.* Briggs & Stratton Corp.

With the upper half of the connecting rod still riding on the crankpin, turn the crankshaft to bring the piston to top dead center (tdc). Force the piston up into the bore with your fingers or with the aid of a wooden dowel pressed against the underside of the piston crown. Lift the piston and rod assembly from the cylinder bore and replace the rod cap, lock, and rod bolts. Observe correct rod-cap orientation and, most particularly, the assembly marks (Fig. 7-18).

Inspect the crankpin and big end rod bearing for scratches, discoloration, and scores. Ideally, bearing surfaces should be perfect; in practice some imperfection will be observed and judgment is required. Most small-engine mechanics accept light scratches and very localized discoloration (no more than 5 percent or so of rod-bearing surface). But deep grooves of the "fingernail-hanging" variety mean that both connecting rod and crankshaft must be replaced.

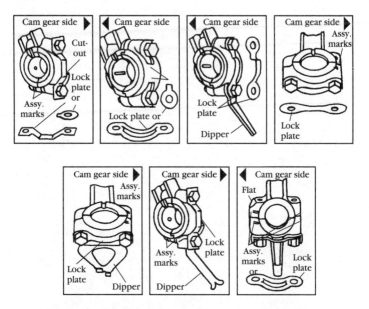

7-18 *Upon disassembly, carefully note connecting-rod cap orientation. Mismatching assembly marks—present on all rods except those with tongue-and-groove interfaces—will destroy rod and crankpin upon startup. Installing the piston-rod assembly 180 degrees out of phase will deny the crankpin lubrication and, in many cases, crash the dipper against the block or camshaft.* Briggs & Stratton Corp.

Big end bearing clearance can be determined with inside and outside micrometers or (more simply and just as accurately) with plastic gauge wire. Wipe the oil off the rod's big end and crankpin and lay a piece of the gauge wire (available from auto parts suppliers) along the length of the crankpin (Fig. 7-19). Without turning the crankshaft, torque rod-cap bolts to specification (Tables 7-3 and 7-4).

Remove the cap and read the bearing clearance in thousandths by comparing the width of the gauge wire to the scale printed on the package. The greater the clearance, the less the wire flattens. Scrape off the remnants of the wire and repeat the operation, this time using two pieces of wire positioned athwart the crankpin (Fig. 7-20). Differences in width between the two wires show crankpin taper; variations in width of the same wire show out-of-round.

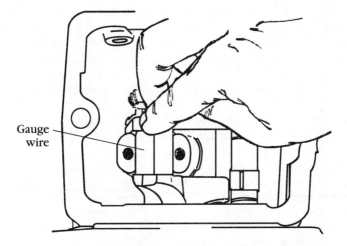

7-19 *Plastic gauge wire placed for running clearance.* Clinton
Engines Corp.

7-20 *Plastic gauge wire placed for taper.* Clinton Engines Corp.

The bearing clearance should fall between 0.0015 and 0.0030
in. Less than 0.0015 in. makes lubrication marginal; more than
0 .0030 in. pounds the life out of the rod and crankshaft, particularly at high speed or under heavy load.

Crankshaft taper is a touchy subject. Some mechanics would
reject a crank that showed any measurable taper. The objection

to taper is that it throws the piston askew in the bore and can, if severe enough, plough the piston pin past the retainer and into the cylinder wall. In practice, one can live with 0.0005 in. of taper if the retainer appears sound and the piston skirt shows a normal wear pattern.

An unusual wear pattern should alert you to the possibility of connecting-rod or crankshaft distortion. A bent connecting rod tilts the piston, leaving a signature that looks like an hourglass (Fig. 7-21). A twisted connecting rod rocks the piston, concentrating wear above and below the piston-pin bores (Fig. 7-22). While connecting-rod twist cannot be verified without an expensive jig, it is possible to detect bends at right angles to the rod bearings. Figure 7-23 shows how this is done using a bar with the same diameter as the piston pin and the block as references. The connecting rod is passed or rejected on the basis of these inspections. No repairs are possible.

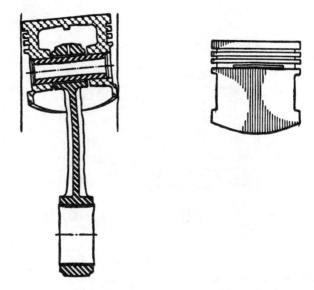

7-21 *A bent connecting rod causes the piston to oscillate in the cylinder, producing the wear pattern shown by the shaded areas.* Sealed Power Corp.

Upon final assembly, wipe all traces of plastic gauge off the rod and crankpin and flood both parts with copious amounts of

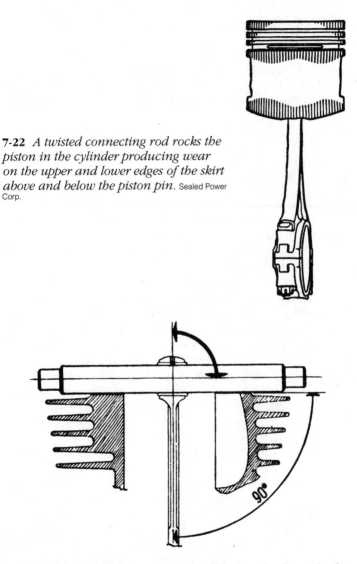

7-22 *A twisted connecting rod rocks the piston in the cylinder producing wear on the upper and lower edges of the skirt above and below the piston pin.* Sealed Power Corp.

7-23 *A rod angularity test, accomplished with the aid of a machinist's square, can be adapted to small engine work.*

engine oil. All friction surfaces must be dripping. Install the rod cap, verifying for one last time that match marks align and that rod-piston orientation is as originally found. Install the rod-bolt lock with tabs straightened and clear of the socket wrench. Pull down the rod bolts evenly and, referring to Table 7-3, torque to specification.

Grasp the rod's big end and attempt to slide it along the length of the crankpin. Some clearance is provided and the rod should move easily under light finger pressure. Turn the crankshaft through several revolutions to detect binding or interference.

Piston

The piston has four functions:

- To react against combustion pressure and convert chemical energy into mechanical motion.
- To hold the rings square in the bore.
- To pass surplus heat out of the chamber and into the cooling fins.
- To isolate the crankcase from the violence of combustion.

The thrust faces—the contact areas on either side of the piston pin and roughly parallel with it—should be lightly burnished. A matted finish, as if the piston had been lapped, means that abrasive particles have accelerated wear. Usually these abrasives are present in the oil, although a failed air cleaner can contribute to the problem. Deep scratches usually mean lubrication failure compounded by overheating. In the most severe cases, the piston welds itself to the bore, depositing splashes of aluminum on the cylinder. A piston with this sort of damage cannot, of course, be reused.

Once the carbon is removed, the piston crown should be smooth and regular, its surface broken only by a suggestion of tool marks. A wavy or flaked piston crown is prima facie evidence of detonation, preignition, or a combination of both.

It's rare, but not unknown, to find a broken Briggs & Stratton piston. The piston might crack at the skirt, in which case it has been loose in the bore and pounded itself to death, or it might crack on the underside at the piston-pin bosses. In either event, discard the piston.

The piston rings change direction twice each revolution. Because the rate of deceleration is more abrupt at tdc, the upperside of the ring groove suffers most. It might appear to be stepped, each step like growth rings on a tree testifying to the increased age of the ring. For as the ring wears, it moves further out in the groove. The upper groove wears four or five times as fast as the second groove. For practical purposes, the third groove, the oil ring groove, does not wear at all.

Remove the rings from the piston and handle them with great care. The edges are razor sharp. Pry one of the upper rings apart, much like you would break a wishbone. The ring should snap cleanly. Secure one of the broken pieces in a file holder and use it as a scraper to remove carbon from the ring grooves. It might be necessary to dull the broken end with a few file strokes to prevent gouging.

Position a new ring in the upper groove and insert a feeler gauge under it as shown in Fig. 7-24. If a 0.006-in. leaf can be bottomed on the base of the groove, the piston should be replaced. The new rings will flutter, fatigue, and break.

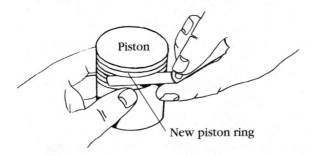

7-24 *A feeler gauge and new compression ring is used to determine piston-groove width.* International Harvester Corp.

Small-engine mechanics rarely measure piston clearance, but depend upon the "wobble test" instead. The piston should have a barely perceptible side-to-side play at tdc (the region of the bore that suffers most). Consideration should be given to replacing the piston if the play is estimated at 0.004 in. More than 0.005 in. calls for a rebore to the next oversize.

A word about replacement pistons. Those used in chromed Kool Bore cylinders are themselves chrome-plated; those intended for iron bores are tin-plated and are further identified by the expander band behind the oil ring and the letter L stamped on the crown (Fig. 7-25). The tin-plated piston cannot tolerate a chromed bore, nor can the chrome-plated piston live in the iron bore.

Since chromed cylinders cannot be replated in the field, oversizing the bores is impractical, and oversized pistons are

Letter "L"

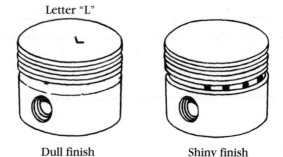

Dull finish   Shiny finish

7-25 *Cast-iron-block engines and aluminum-block engines with sleeved bores require a tin-plated piston, identified by the dull finish and the letter L stamped on the crown. Kool Bore (chromium-plated, aluminum-bore) engines use a bright piston without the identifying mark.* Briggs & Stratton Corp.

not available. Iron-bore pistons are available in oversizes of 0.010, 0.020, and 0.030 in. The oversize is stamped on the crown.

Before separating the piston from the connecting rod, note any reference marks on the piston. Some pistons have the letter F embossed on the skirt. Others have notched crowns. These reference marks should be on the flywheel side. Make note of the connecting-rod match marks, referencing them to the camshaft or some other prominent feature. Otherwise it is possible to reassemble the shank wrong. Solid piston pins have a recess on one side and should be assembled as found originally.

Using long-nosed pliers, remove both piston-pin retainer clips and discard them. Retainers are too important to be trusted a second time around. Remove the piston pin. Figure 7-26 shows a tool of the type supplied by motorcycle and snowmobile manufacturers. Another method is to heat the piston by wrapping it in rags soaked with hot oil, or by placing the piston crown down on an electric hot plate. Yet another method is to support the piston in a wooden V-block and drive out the pin with a punch. Be sure to keep the punch against the pin because if it wanders into the bearings the piston will be ruined.

Piston-to-pin clearance is 0.0005 in. for all engine models. Rejectable connecting-rod pin-bore sizes are provided in Tables 7-3 and 7-4. Oversized (0.005-in.) piston pins are available for use in worn engines.

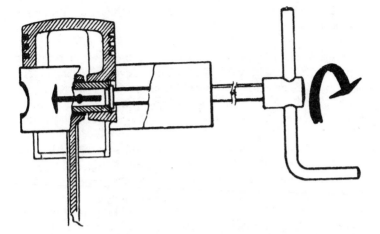

7-26 *A piston pin press is relatively easy to construct.* cz

Assemble with new piston-pin retainer clips. Place a retainer clip in its groove in one piston-pin bore. Lubricate pin, piston, and connecting bearings. Insert the piston pin from the opposite side of the piston, flat end first for solid pins, either end first for hollow pins. Press the pin through the nearside bore, mate it with the connecting rod (observing correct rod and piston orientation), and press the pin home against the retainer clip previously installed. Install the remaining retainer clip and verify that both retainer clips are seated in their grooves.

Piston rings

Counting from the top, the purpose of the first and second ring is to seal compression and combustion pressure. In addition to its primary function, the second ring scrapes surplus oil from the bore, and is sometimes called the scraper ring to distinguish it from the uppermost, or compression ring. The bottom, or oil-control ring, lubricates the cylinder.

A piston ring is a pressure-compensating seal. When there is no pressure in the cylinder, the ring lies dormant, exerting only a few ounces of residual spring tension against the bore. As pressure above the ring rises, some of this pressure bleeds over the upper edge of the ring and, acting from behind it, cams the ring hard against the bore. The greater the pressure, the stronger the camming action, and the more tightly the ring hugs the bore.

The ring must have some residual tension, otherwise the pressure escapes past the ring face. Severely worn rings, or rings that have been overheated, cannot develop the initial tension. By the same token, the ring must be free to move in the bore. Carbon- and varnish-bound rings have ceased to function; whatever compression the engine has is developed by the piston.

Figure 7-27 illustrates five "factory" ring sets. Note that the compression and scraper rings have a definite top and bottom. Installing these rings upside down costs compression and, in the case of the scraper, increases oil consumption. The upperside is identified by the words *top* or *up*. Oil-control rings and expanders (not shown) are symmetrical, and can be installed either side up.

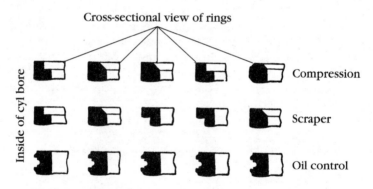

7-27 *Cross-section of various Briggs & Stratton rings. Unshaded areas are inboard next to the piston.*

Briggs & Stratton also supplies "engineered" ring sets for use in severely worn iron bores. These chrome-plated rings tolerate 0.005-in. cylinder wear and seat themselves without the need for honing.

The ring gap—the installed distance between the ring ends—is the primary measure of ring wear (Fig. 7-28). The more the ring faces have eroded, the wider the gap. In this context, the wear limit is 0. 035 in. on both compression rings and 0.045 in. for the oil-control ring for aluminum Kool Bore applications. Iron-sleeved engines are set up a little tighter, with ring wear

7-28 *Determining the ring-end gap in a Kohler engine. The same technique can be used to give some approximation of bore wear and taper.*

limits of 0.030 in. for the compression rings and 0.035 in. for the oil-control ring.

The variation in ring gap at different positions in the bore is a poor man's cylinder gauge. Establish the gap at the base of the bore in the region below ring travel. This gap represents zero bore wear. Each increase in gap above this null point translates as cylinder wear, and/or eccentricity. The change from the bottom to the zone just under the ridge is the cylinder taper. When making these measurements, use a piston as a ram to keep the ring square.

Ring gap is also a way to detect manufacturers' mistakes. It is unusual, but by no means spectacularly rare, for rings to be mislabeled. Determine the gap of each new ring. This gap should be considerably less than the wear limit, certainly no more than 0.0015 in. per inch of bore diameter.

The big problem with rings is hardly mentioned in the literature, and rarely acknowledged by mechanics. Rings have a tendency to break *after* overhaul, within a few hours of startup.

No one is surprised to find broken rings in a worn engine; indeed, the fact that one or more rings has shattered justifies the mechanic's work. But replacement rings should be immune to breakage. In some cases, particularly those involving the upper compression ring, an overly wide ring groove is the culprit. But post-overhaul breakage occurs about as often when rings are fitted on a new piston, a piston that should have correctly sized grooves. Break-in stresses can hardly be blamed, since factory-new engines seem immune to early ring failure.

The reason must be in the way the mechanic handles the rings while they are in his charge. He performs two operations: expanding the rings to slide over the piston, and compressing them to enter the cylinder bore. If done incorrectly, either of these operations can weaken the rings.

Purchase or rent an expander like the one shown in Fig. 7-29. This tool pries the ring open while holding the ends on the same plane. Install the oil-control ring, then the scraper, and finally the compression ring. Some oil-control rings are backed by an expander spring that may be installed with either side up. Check that the upper two have their correct sides up and rotate each ring 120 degrees from the others. If the gaps were to be in line, the rings might "freeze" in that position, leaking compression and exhaust gases on each upward stroke. Staggered, the rings are free to rotate, which discourages groove sticking, and are extremely unlikely to align their gaps again.

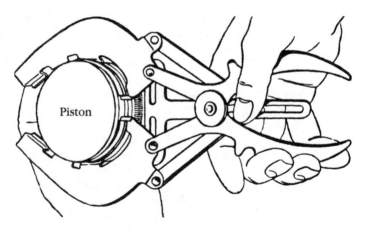

7-29 *A piston-ring expander is used to remove and install piston rings.* Briggs & Stratton Corp.

Lubricate the rings, piston pin, and piston skirt with high-grade motor oil. The old method, now in disfavor because of the spark-plug fouling it caused, at least was thorough. The mechanic simply immersed the piston, crown down and pin deep, in a bucket of oil.

Ring compressors for small engines are available from K-D Manufacturing, Whitney, and of course, from Briggs & Stratton and its competitors. Mount the piston assembly in a vise with blocks of hardwood between the jaws to protect the connecting rod. Tighten the compressor over the rings, exerting just enough pressure to overcome residual ring tension. It is not necessary to bear down hard on the compressor. Place the assembly over the bore, aligning any index marks on the piston or rod, and push the piston out of the tool and into the cylinder (Fig. 7-30). Thumb pressure should be adequate. Do not force the issue. If the piston hangs, a ring has escaped the confines of the compressor, or the rod is jammed against the crankshaft. Withdraw the assembly and start over.

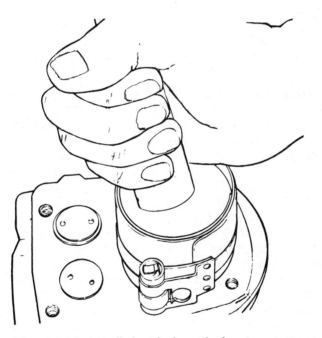

7-30 *The piston is installed with the aid of a ring compressor, shown here being used on a Kohler engine.*

Once the piston is in the cylinder, torque up the rod, and turn the crankshaft a few revolutions. The piston should move without protest, slowing a bit as it transverses the middle of the stroke where piston movement is greater per degree of crankshaft rotation than at the dead centers.

Cylinder bore

One of the mechanic's responsibilities is to gauge the cylinder bore at three places along its length: at the upper limit, center, and lower limit of ring travel. Make two measurements at each station, one in line with the piston-pin axis, and one at right angles to it.

Bore diameter should be measured with a direct-reading cylinder gauge or with an inside micrometer. Most small engine mechanics estimate bore wear with the wobble test and the fortitude of their convictions. The wear limits of the cylinder bore are:

All models	0.0035-in. oversize
Cast-iron blocks	0.0015-in. out-of-round
Aluminum blocks (iron-sleeved)	0.0025-in. out-of-round

Boring   Cast-iron blocks and aluminum blocks with iron sleeves can be bored 0.010, 0.020, and 0.030 in. past stock diameters. Cast-iron blocks will accept larger overbores, when a suitable piston can be found.

There are three ways to bore the block. Assuming first-rate tooling and reasonable competency on the part of the machinist, the best method is to use a Sunnen or equivalent automatic hone. The next best choice is to cut the bore in a lathe. The least desirable method is to hone the cylinder oversize by hand.

Since machine tools are expensive and few customers come in for major engine work, most shops use a hone. An adjustable hone, such as the one illustrated, gives better accuracy than the conventional, spring-loaded type. Select the stone from the hone manufacturers' recommendation, remembering that grit size (the number of abrasive particles per square inch) and hone code numbers are not always identical. For example, a number-500 stone is equivalent to 280 grit.

Some mechanics prefer to run the hone dry, but once the stones are wetted, they must continue to be lubricated. Kerosene and other petroleum oils dissolve the adhesive that binds the particles in the stone and cause rapid wear without a corresponding increase in cutting speed. Use a commercial honing oil, animal fat, or vegetable oil such as Crisco.

Although cylinders have been bored with a ¼-in. drill motor, a drill press is the better choice. The easiest way to secure the block to the drill-press table is with a heavy work plate and clamp as shown in Fig. 7-31A. Most blocks must be shimmed to level the fire deck. Another, more professional, setup involves construction of a honing plate that makes up to the flange-bolt pattern on all engine models (Fig. 7-31B). The next figure (Fig. 7-32) is a working drawing of the plate. Set the plate and block on a well-oiled drill-press table so that the assembly is free to align itself with the hone. Adjust the spindle stop to limit hone travel to ¾ of an inch past either end of the bore, and gear the machine down to 400 rpm. Reciprocate the hone to about 70 strokes a minute. If you use a spring-loading hone, concentrate on the lower part of the bore and slowly increase the length of the strokes to include the whole bore. Sweep the whole bore

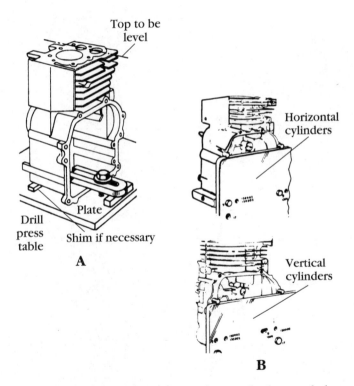

7-31 *Secure the block of honing with a work plate and clamp (A) or construct a honing plate (B).* Briggs & Stratton Corp.

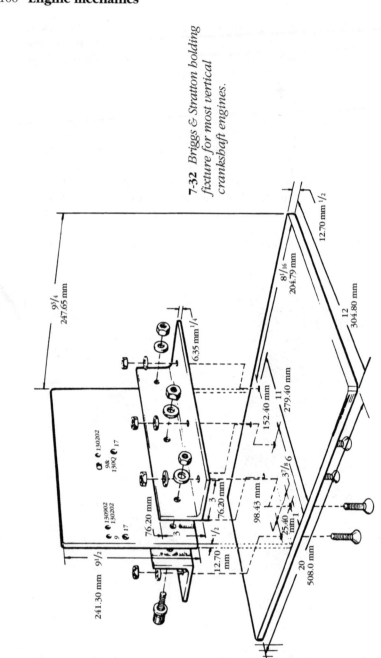

7-32 Briggs & Stratton holding fixture for most vertical crankshaft engines.

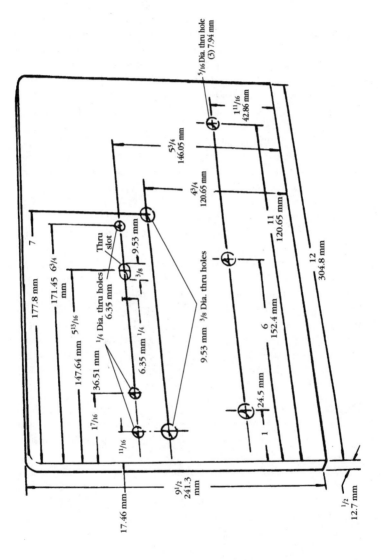

with an adjustable hone, opening the stones until the cut is uniform from top to bottom.

Briggs & Stratton piston oversizes are based on the diameter of the original piston, and not on bore diameter. In other words, a piston marked 0.010 is ten-thousandths of an inch larger than the factory-standard piston. If the cylinder bore is enlarged 0.010 in., this piston will have the correct running clearance. A machinist might be confused on this point since other manufacturers base piston size on bore. In this system, a piston that is marked 0.010 is exactly ten-thousandths of an inch larger than the original bore. The bore must be cut out further for piston clearance.

Whether the bore is machined or honed, the last 0.0015 in. or so should be finish honed with a fairly light stone. Otherwise the rings are slow to seat and might never give adequate oil control. Cast-iron rings seat best with a number (not grit number) 200 stone; chrome rings need a 300 stone; and stainless-steel rings require the almost polished finish given by a 500 stone. Check work with an accurate micrometer after the cylinder cools.

Abrasive particles must be removed from the bore if the honing process is not to continue over the life of the engine. Petroleum-based solvents merely float the abrasive deeper into the pores of the metal. The only way to clean a honed bore is with hot water, detergent, and a sturdy brush. Scrub the bore thoroughly and wipe it down with paper towels. If the towels discolor, abrasive is still present and more scrubbing is required. Once the bore is completely free of abrasive, oil it to prevent rust.

Deglazing   Cast iron is an amazing material, used in its heyday for everything from lamp posts to structural beams. The Crystal Palace, symbol of Victoria's reign and British industrialism, was framed in cast iron.

One of this metal's unique properties is the way it forms a glaze when subject to rubbing friction. The working surface flows and compresses into a smooth, hard skin that is almost impervious to further wear. Without the formation of glaze, it would be impossible to run cast-iron rings in iron bores.

Glaze is beneficial to an older engine, but it is a nuisance when the engine has been overhauled and the rings replaced. Its imperviousness to wear means that the rings do break in properly. Break-in is a mutual honing process, a period of accelerated ring and bore wear until the parts accommodate themselves to each other.

Deglazing, or "glaze busting," is no more than light honing for the purpose of roughening the metal and giving it tooth. Done correctly, the cylinder wall takes on a dull, matted appearance. Upon microscopic examination, the surface is characterized by diamond shaped peaks and valleys. The peaks are in rubbing contact with the piston rings, while the valleys become oil channels to cool the rings and float away the debris.

Use the appropriate finishing stone for the ring material as described in **Boring**, and turn the hone no more than 400 rpm. Reciprocate it at approximately 70 strokes a minute. The cylinder might be said to be deglazed when the surface is entirely honed, although some discretion should be used in the matter. The hone will skip in an eccentric, or egg-shaped, cylinder. Cylinders with severe eccentricity should either be overbored to the next size, or left alone. Taking out the oval costs more metal than it is worth.

Deglazing is inappropriate for chromed bores that do not develop a "skin" and for cast-iron bores fitted with the Briggs & Stratton-engineered ring set.

Crankshaft

The crankshaft is the fundamental part upon which all else depends. Because it is a passive element, having nothing directly to do with performance, mechanics sometimes take the crankshaft for granted. But flaws in the crankshaft mean that the engine will not live, and all of the polishing and fitting amounts to nought.

Begin the inspection with the flywheel threads (Fig. 7-33). Pulled or crossed threads mean that the flywheel can break free. The keyway must be square and sized to fit the key, otherwise the magneto will jump time and starting will become problematic. Bearing surfaces should be reasonably close to specification, and smooth (small imperfections can be polished out, but deep, nail-hanging ridges mean the crankshaft must be replaced). Without a micrometer, you can get some idea of main journal wear by moving the crankshaft from side to side in the bearings. There should be no, or almost no, play. The crankpin can be checked with a new connecting rod and plastic wire gauge. The crankshaft should be straight to within

Discard crankshaft if small or out of round

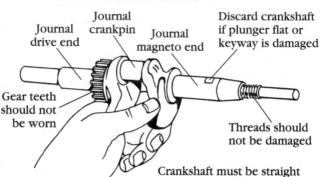

7-33 *Crankshaft inspection points.* Briggs & Stratton Corp.

0.001 in., a determination that is made with a dial indicator and a set of precision V-blocks.

Main bearings

Briggs & Stratton engines use plain or ball main bearings. When ball bearings are specified, the inner race makes a press fit with the crankshaft, and the outer race is pinned to the bearing carrier that, in turn, is bolted to the block. Plain bearings take the form of a bushing on cast-iron engines, and are integral on most aluminum blocks. A few models use replaceable bushings.

Ball-bearing clearances cannot be gauged. The condition of the bearing is judged by how it looks and sounds. Soak the bearing in solvent to remove all lubricant and allow it to air dry. If you use compressed air, the source must be fitted with a water trap, since antifriction bearings have zero tolerance for moisture. Turn the bearing by hand. It should roll without catching and without the cracks and pops associated with brinelled faces and pitted balls. Some outer-race play is allowable, although it should be limited to a few thousandths of an inch.

If the bearings have failed, remove them from the crankshaft with an arbor press or a bearing splitter (Fig. 7-34). Drive new bearings home with a length of pipe sized to match the diameter of the *inner* race (Fig. 7-35). A more refined method is to warm the replacement bearings in a container of oil. Maximum allowable oil temperature is 325 degrees Fahrenheit, and the bearings must be held clear of the bottom and sides of the con-

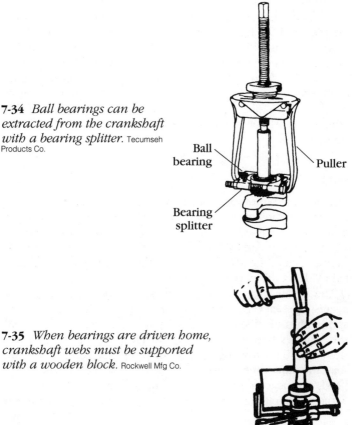

7-34 *Ball bearings can be extracted from the crankshaft with a bearing splitter.* Tecumseh Products Co.

Ball bearing

Puller

Bearing splitter

7-35 *When bearings are driven home, crankshaft webs must be supported with a wooden block.* Rockwell Mfg Co.

tainer. Once warmed, the inner races will expand enough to slip over the crankshaft. The shielded sides of the balls are inside, toward the crankpin.

Magneto-side plain bearings can be renewed on all engines. Sump-side bearings can be renewed except on Models 9, 14, 19, 20, and 23 in the cast-iron series; 8B-HA, 80590, 81590, 80790, 81790, 82990, 92590, 92990, 110990, and 111990 in the aluminum series. PTO-side reamers are not available for these engines, and new oil sumps or bearing covers must be purchased.

Magneto-end seal

The crankcase, or magneto-end, seal is removed and replaced as described previously in this chapter in **Flange**. Figure 7-9 illustrates the extraction process. Install the new seal with the steep side of the lip toward the sump and drive it to the depth of the original with a soft wooden block or, better, a length of tubing turned to accurately match crankcase seal-bore inner diameter (ID). If the crankshaft is grooved, the seal sometimes can be repositioned a fraction of an inch to compensate. However, the back of the seal must make good contact with the bore and must not, at the other extreme of travel, invade the oil port.

Valves

The primary symptom of valve failure is loss of compression. If the engine runs at all, power will be drastically curtailed, although engine speed will be unaffected. Intake-valve failure is often accompanied by "pop-back" through the carburetor bore. In this regard, a carbon-blackened throttle plate is circumstantial evidence of intake-valve problems.

The whole purpose of the valve is to form a gas-tight seal against the seat (Fig. 7-36). For the seal to be effective, the valve spring must exert sufficient tension to compensate for the small irregularities between the seat and face, the valve guide and stem must be within tolerance to hold the face concentric to the seat, the valve stem must be straight and at right angles to the head, the head must be round, and seat and face must mate without the interference of impacted carbon or displaced metal.

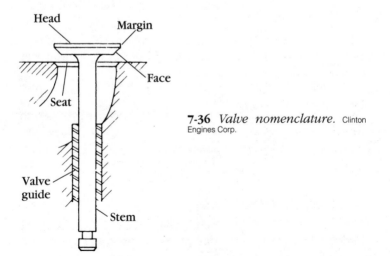

7-36 *Valve nomenclature.* Clinton Engines Corp.

Intake valves are susceptible to all of these failures and have an unfortunate affinity to attract deposits under their head. Coking, or carbon accumulation, is usually present in lightly loaded engines. Gum deposits are the mark of a poorly maintained engine, one that has been allowed to run chronically rich. Gum deposits can also be caused by using stale gasoline, and by habitually shutting the engine down before it has reached operating temperature.

The exhaust valve is most susceptible to these maladies and usually fails first. Under-head deposits might be brown, yellowish, or tan, in which case they are normal byproducts of combustion. Stem corrosion is not normal, and is caused by moisture in the fuel. It also occurs when the engine is repeatedly stopped before it attains operating temperature. Overheating is signaled by a black, polished appearance and the relative freedom from deposits. Suspect that the stem is bent.

The valves are secured by keepers, which in turn are held by valve-spring tension. Three types of keeper have been used, the late and unlamented pin, the one-piece retainer, and the split collet. The pin passes through a hole in the valve stem and bears against a collar. As the engine ages, the hole elongates and the pin bends. If your budget allows, or if events demand new valves, purchase the type with slotted stems. Use one-piece retainers (Fig. 7-37) or split collets (Fig. 7-38). Split collets are preferred for modified engines.

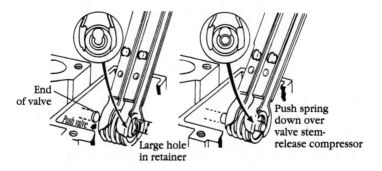

7-37 *Installing Briggs & Stratton one-piece valve retainers.*

Both illustrations show Briggs & Stratton PN 19063, probably the finest valve-spring compressor available for small engines. Without this tool, you can make do with a pair of screwdrivers,

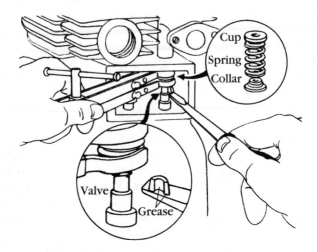

7-38 *Installing split collets.* Briggs & Stratton Corp.

but the work is awkward. Apply pressure to both sides of the collar (or to one of the lower coils if no collar is fitted) to raise the spring without lifting the valve. Have a helper disengage the keeper. Reverse the procedure when installing the valves.

Valve springs

Briggs & Stratton does not publish valve-spring tension specifications, and one must fall back on the old mechanic's rule that the free-standing height of a used spring should be at least 90 percent of the height of a new spring. The spring should stand perpendicular. Tilt means that the coils have weakened. Examine the spring for pitting (the early symptom of fatigue) and for fretting (the sign of coil binding). Replace as necessary.

Valve guides

Many Briggs & Stratton engines are without valve guides as such. They run the valves directly against block metal. In defense of this practice, it should be remembered that the valves operate in a vertical plane, without the horizontal forces generated by rocker arms. White metal and cast iron are adequate support, at least through the first overhaul. But insofar as longevity is concerned, nothing substitutes for an honest bronze guide.

The traditional (and fairly accurate) test is to raise the valve to its full extension and move it side-to-side. A wobble of ¹⁄₁₆ of an inch or more means the guide has bell-mouthed and should be knurled or replaced. Another method is to use a factory gauge, PN 19122 for engines displacing 13 cubic inches or less (excepting Model 9), and PN 19151 for the others. If the gauge can be inserted ⁵⁄₁₆ of an inch or deeper into the guide, it is worn.

Briggs & Stratton engineers would have the guide reamed oversize and a bushing pressed into place. For smaller engines (¼ -in. guide ID), drawing no more than 13 cubic inches (except the Model 9 with its large valve stems), the necessary tools are:

Part number	Name
19064	Reamer
19191	Reamer Guide Bushing
19065	Bushing Driver
19066	Finish Reamer

Centering the reamer with the guide bushing, ream the valve guide ¹⁄₁₆ of an inch deeper than the length of the replacement bushing. Do not ream the guide completely through, for its lower diameter is needed to stabilize the bushing. Drive the bushing flush with the top of the valve guide using the factory tool or a copper punch. Use the finish reamer to size the bushing to the valve stem. Both reamers are turned clockwise and lubricated with cutting oil or kerosene.

Engines displacing 14 cubic inches and more, as well as the ubiquitous Model 9, use ⁵⁄₁₆-in. ID guides.

Part number	Name
19231	Reamer
19234	Reamer Guide Bushing
19204	Bushing Driver
19233	Finish Reamer

As described in the previous paragraph, ream the valve guide ¹⁄₁₆ of an inch deeper than the replacement bushing. Lubricate the guide ID with PN 93963. Be careful that this product doesn't get on the tappets. Use a soft copper punch to drive the bushing flush with the top of the guide. Since bushings are finish reamed before shipment, no reaming tool is supplied.

Sintered iron, brass, and aluminum guides are extracted with a tap driven into place and reamed to size.

Valve and seat refinishing

Once you are satisfied that the valves are basically sound, that the springs generate enough tension to close the valves, and that the guides hold the valves concentric to the seats, the next operation is to refurbish the valve faces and seats. Unless these prior conditions are met, grinding is a waste of time.

Lapping is a short-term fix, an expedient to get the engine back into service until the valves and seats can be properly machined. For lapping to be at all effective, the valve faces and seats must be relatively healthy. Deep pits require heavy lapping, widening the seating surface and possibly grooving the valve face. A wide seating surface tends to collect carbon and to develop less unit-area pressure than a narrower seat. The grooving effect becomes serious when the engine reaches operating temperature and the parts "grow." The engine suffers a hard to diagnose loss of power (Fig. 7-39) when what was a good seal at room temperature suddenly evaporates.

The most popular lapping compound is Clover Leaf brand. An oil-based mixture, it is sold in double-ended tins. One compartment contains a coarse abrasive for rapid cutting, the other a fine finishing compound. Unless you are dealing with an antique like the Model N, you will need some way to turn the valves. K-D markets a small suction cup tool designed for small engines. It is available through most auto-supply houses as catalog number 501. But suction cups tend to slip on highly polished valve heads, and it might be necessary to moor the tool with a drop of contact cement.

Apply several dabs of coarse compound to the valve face with a screwdriver, being careful not to overdo it. Surplus compound might find its way to the valve guide where it would be ruinous. Hold the valve by the suction cup and rotate it back and forth. Every four or five seconds, give the valve a half turn so that the whole surface will be lapped evenly.

Renew the compound at frequent intervals, and whenever you no longer hear the harsh grating noise that accompanies lapping. Running the valve dry scores the seat and face. Finish the job with fine compound, and clean the parts with a rag dipped in solvent.

Valve grinding involves the use of a valve lathe and a high-speed seat grinder. While this equipment is expensive for the amateur or occasional mechanic to purchase, the work can be farmed out to an automotive machine shop for a few dollars. The benefits are considerable. First, since the lathe centers on

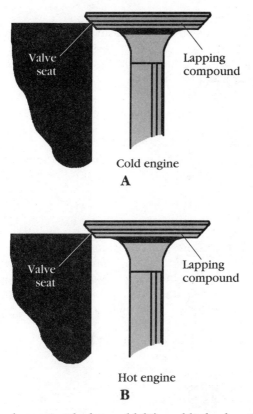

7-39 *A valve can seal when cold (A) and leak when the engine warms and the valve stem expands (B).* Clinton Engines Corp.

the valve stem and the grinder on the valve guide, both parts are concentric. Small irregularities, a slightly bent stem or a canted valve guide, are compensated for automatically. Second, the valve face and seat are cut at uniform angles, and hence, heat expansion has little effect upon the seal. Third, the seat can be narrowed to factory specifications by using alternate stones. And finally, if the work has been done correctly, no lapping is necessary.

The valve-face angle for current Briggs & Stratton production is 45 degrees on both the intake and exhaust. The seats are ground at the same angle, without the half a degree of "interference fit" favored by some other manufacturers. Early Briggs engines had intake valves and seats ground at 30 degrees for better flow. The valve margin should be at least ¼₄ of an inch thick to control valve temperatures and reduce the possibility of preignition (Fig. 7-40). While it is possible to widen the margin by judicious grinding, the valve should be replaced. Recommended seat width is ¾₄ to ¼₆ in.

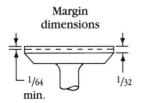

Margin dimensions

1/64 min.   1/32

7-40 *Valve margin dimensions.* Clinton Engines Corp.

Once the valves are lapped or ground, the stems must be shortened to compensate for the metal removed. Place the valves in their guides, but do not install the springs and keepers. Turn the flywheel until one valve is fully open. Turn the flywheel one more complete revolution and adjust the clearance between that valve and the tappet with a feeler gauge (Fig. 7-41). See Tables 7-3 and 7-4 for specifications.

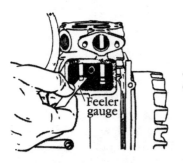

Feeler gauge

7-41 *Determining valve lash.* Clinton Engines Corp.

To adjust the clearance, grind the stems flat, "kissing" the wheel to remove a few thousandths of an inch of metal. It is disconcertingly easy to remove too much metal, so check and finish the operation with a file, slightly breaking the square edges. Then repeat the process for the other valve.

Some 170000 and 190000 and all Model 220400, 243000, 300000, 250000, 320000, and 420000 engines feature Cobalite exhaust valves and positive valve rotation in the form of a cammed device called a Rotocap. These parts, illustrated in Fig. 7-42, extend exhaust valve life. Models 170000 and 190000 *might* be supplied with Cobalite exhaust valves—usually identified by the letters TXS on the head—and are *always* fitted with a Rotocap (Fig. 7-42). And to add a little more confusion, all Briggs & Stratton engines can be ordered with Cobalite exhaust valves. But no engine intended for LPG (liquefied petroleum gas) operation ever left the factory with a Rotocap. LPG is a dry fuel, which is unfortunate from the point of view of the valves that depend upon lead in regular gasoline for lubrication. Positive valve rotation would quickly score the valve seat.

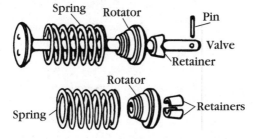

7-42 *Stellite exhaust valves use pin or split collet retainers.* Briggs & Stratton Corp.

Valve seats

The intake valves on cast-iron engines run directly on the block. Fortunately, cast iron is a relatively good seat material. Exhaust valves on iron engines and both valves on aluminum models run on replaceable seats.

Valve seats might crack, wear thin with age, or work loose from the block. In most cases, port geometry is such that the seat can be driven out from below with a punch. Other engines require a valve-seat puller, available from Briggs & Stratton in four varieties to fit the various models (Fig. 7-43). A substitute can be fabricated easily. Replacement intake-valve seats are inventoried for cast-iron engines, although the cost of the counterbore reamer and pilot is prohibitive for an occasional mechanic. It's better to farm this job out to a Briggs & Stratton repair station.

Once the old seat is removed, clean all traces of oil from the recess in the block. See that it is true, flat, and without grooves

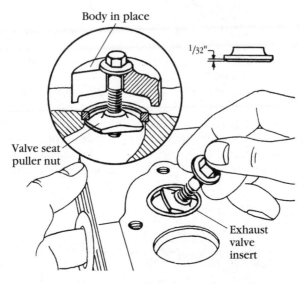

7-43 *Valve-seat inserts should be peened on aluminum-block engines.* Briggs & Stratton Corp.

or other imperfections that would deny a perfect fit between the seat insert and block.

Most mechanics simply drive the seat home using special tool PN 19136 and pilot PN 19126 on engines displacing less than 14 cubic inches, and pilot PN 19127 on larger engines. And some mechanics simply use an old valve as the driver. A more civilized method than either is to heat the block for several hours in the kitchen oven at about 275 derees Fahrenheit. Support the casting on bricks to prevent local overheating. Chill the seat insert in the freezer or by packing it in crushed dry ice and alcohol. Working quickly before the parts normalize, drop the seat into the recess with the beveled edge up. The top of the seat should be approximately flush with the firedeck on cast-iron engines and a few thousandths of an inch below it on aluminum blocks. Since the thermal expansion rate of aluminum is four times that of cast iron, the seats on these engines should be peened around their full circumference (Fig. 7-44).

Crankcase breather

The crankcase breather vents corrosive gases from the crankcase and, at the same time, maintains the crankcase at a slight

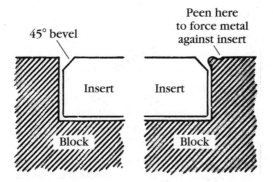

45° bevel

Peen here
to force metal
against insert

7-44 *Peening the valve seat secures it to the block.* Briggs & Stratton Corp.

negative pressure. The breather assembly doubles as the valve-chamber cover and is secured by two fillister- head screws. The breather usually vents to the carburetor, an arrangement that explains why a poorly maintained Briggs & Stratton will sometimes run even when the fuel tank is empty. There is enough raw fuel in the oil to support combustion. The symptoms of breather failure are:

- Oil leaks at the flange gasket, crankshaft seals, or through the breather.
- Rapid oil discoloration and sludging.

Wash the assembly in solvent, let it drain, and check the valve clearance. The fiber disc should clear the housing by 0.045 in. (Fig. 7-45). Bend the bracket as necessary.

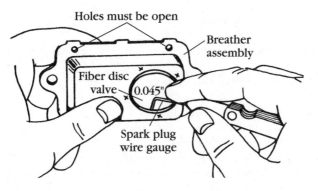

7-45 *Checking the crankshaft breather assembly.* Briggs & Stratton Corp.

Two-cycle notes

The two-cycle breaks with company tradition in several ways, not the least of which is its operating cycle. With the exception of the 62032 snow-blower engine, all other Briggs products operate on the four-cycle principle.

The now-defunct 95700 had a chrome-over-aluminum bore; the current 96700 uses a conventional iron liner, which should probably be retrofitted to the older engine at rebuild time.

Engine accessories—carburetor, rewind starter, mechanical governor, muffler, and ignition system—are standard Briggs fare, modified slightly for the application. But mechanics unfamiliar with Japanese and European design practices will find much to ponder in this metric engine.

Four Allen screws secure the one-piece cylinder barrel and head to the crankcase. Piston rings straddle pins that keep the ring ends clear of the exhaust port. Because of interference fits at the main bearings, special tools are required for disassembly and assembly. Hirsh-type built-up crankshafts support one-piece connecting rods. Crankshafts and attached connector rods are serviced as assemblies.

Piston & rings

Using a 5-mm Allen wrench, remove the four socket-head screws securing the barrel to the crankcase assembly (Fig. 7-46). Break the gasket bond with a rubber mallet.

Support the piston with a wooden block, notched to fit over the connecting-rod shank (Fig. 7-47). Without this holding fixture, inadvertent crankshaft movement would bang the piston skirt against the crankcase. Stuff a rag into the crankcase cavity to keep piston-pin locks and other foreign objects out.

With an ice pick or similar tool, pry and peel off one piston-pin lock.

Warning: Wear eye protection when removing and installing pin locks.

Fabricate the tool illustrated in Fig. 7-48 and, using palm force only, push out the piston pin. Remove the remaining pin lock. Figure 7-49 illustrates the parts arrangement. Two-cycle piston pins and pin bearings do not seem as durable as their four-stroke counterparts, perhaps because of the difference in load profiles. Table 7-5 lists pin-bore and pin-reject

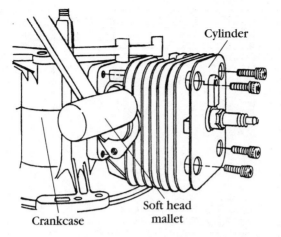

7-46 *Briggs & Stratton 95700 and 96700 engines employ a European-style one-piece cylinder. A rubber hammer might be required to break the barrel/crankcase-gasket seal.*

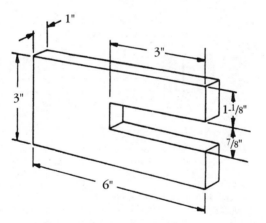

7-47 *Dimensions for the piston fixture, constructed of wood. Figure 7-53 shows this tool in use.* Briggs & Stratton Corp.

dimensions. Replace the bearing as insurance and always when a new pin is fitted.

Remove the two rings with a ring expander (shown in Fig. 7-29) and introduce one ring into the cylinder barrel. The cham-

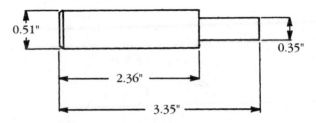

7-48 *Piston-pin tool, also constructed of wood.* Briggs & Stratton Corp.

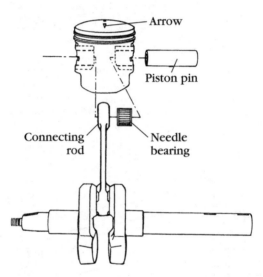

7-49 *Piston assembly, showing the pin, pin bearing, and orientation. The arrow goes toward the exhaust port.* Briggs & Stratton Corp.

ber acts as a ring compressor. Using the piston as a pilot, push the ring about 1¼ in. into the bore and measure the end gap to gain some idea of engine condition. The 0.040-in. wear limit is a bit academic, since few mechanics would disassemble a two-cycle engine this far without replacing the rings.

Deep, fingernail-hanging score marks, damage to the ring lands, stress cracks, or severe abrasion mean that the piston has reached the end of its useful life. If there are no obvious defects, measure the extent of wear at the thrust faces (900 degrees to the pin bosses) as shown in Fig. 7-50. The wear limit is 2.356 in. or less for both engine models.

Table 7-5.
Two-cycle (95700 and 96700) specifications

Compression	90–120 psi (during cranking @ wide-open throttle)
Armature air gap	0.008–0.015 in.
Crankshaft endplay	0.002–0.012 in.
Wear limits:	
Cast-iron cylinder bore	2.368 in. or greater
Chrome cylinder bore	2.369 in. or greater
Piston ring gap	0.040 in. or more
Piston (across thrust faces)	2.356 in. or less
Piston pin bore	0.553 in. or greater
Piston pin OD	0.550 in. or less
Crankshaft main bearing journals	0.982 in. or less
Torque limits:	
Spark plug	160 lb/in.
Flywheel nut	30 lb/ft.
Cylinder mtg. screws	110 lb/in.
Crankcase screws	60 lb/in.
Carburetor mtg. bolts	50 lb/in.
Muffler mtg. bolts	85 lb/in.

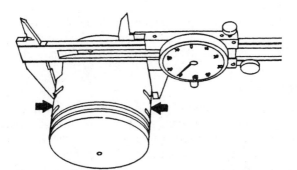

7-50 *Measure piston diameter at the point indicated.* Briggs & Stratton Corp.

Note: Most piston defects mirror defects in the cylinder bore.

Use a wooden spatula or some other forgiving tool and scrape carbon deposits from the combustion chamber, ports, and piston crown. The factory warns against using steel scrapers and wire brushes, although it is difficult to imagine that a brass brush would do much harm.

Clean piston grooves with a broken ring, as described in the previous **Piston** section for four-cycle engines. Install new rings oriented as shown in Fig. 7-51. The upper ring carries the letter N on its top side; the lower ring has no distinguishing marks and can be installed either side up. Center the ring ends over the locating pins.

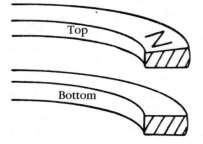

7-51 *No. 1 compression ring installs with its angled side up—identified by the letter N. No. 2 has no directional orientation.* Briggs & Stratton Corp.

Make up the piston and small-end bearing to the pin and connecting rod with the tool fabricated in Fig. 7-48. Lubricate the pin, piston-pin bores, and needle bearing. The arrow on the piston crown points toward the exhaust port.

Caution: Failure to orient the piston correctly might result in ring breakage.

Install new lock rings, oriented with the open ends on the cylinder-bore centerline—with open ends toward the top of the barrel or else turned 180 degrees and pointing straight down.

Cylinder

Inspect the cylinder bore for obvious defects, such as scoring, aluminum smears, and damage to the chrome plating (Model 95700). Using a telescoping gauge, measure cylinder diameter at the top, middle, and bottom of ring travel. Make two measurements at each station, 90 degrees apart, to detect possible out of round that, if standard Briggs specs apply, should not exceed 0.0015 in. for iron and 0.0025 in. for chrome. The iron bore can be worn to 2.369 in. and the chrome/aluminum bore to 2.368 in.

before cylinder replacement becomes mandatory. The 0.001-in. difference is conditioned by the thickness of the plating.

No purpose is served by honing chromed bores. Used iron cylinders might benefit from "glaze breaking," but few mechanics bother. Two-cycle rings do not need incentive to accelerate wear.

Lower end

Servicing crankcase components is a fairly complex business that requires a collection of special tools, packaged by the factory under PN 19332 (Fig. 7-52). A mechanic who has only occasional contact with these engines could wing it with homemade pullers and help in the bearing department from a competent automotive machinist.

When pressing in bearings, support the underside of the crankcase castings at the bearing boss and not on the casting perimeter. Failure to observe that little nicety will result in a "tin-canned" casting. Some relief can be had by warming the castings in hot oil (main-bearing oil seals limit thermal input) and packing the bearings in dry ice. But at some point, you are going to have to use press force and will need the proper driver.

In basic outline, the work goes as follows:

1. Remove the four screws that secure the crankcase halves.
2. With a puller taking purchase in the screw holes just vacated and reacting against the crankshaft end, pull the lower crankcase half off its bearing.
3. Disengage the upper, or magneto-side, casting from the crankshaft. This operation, illustrated in Fig. 7-53, requires the same tooling used in Step 2.
4. Replace the crankshaft and integral connecting rod if main-bearing journals measure 0.982 in. or less, or if 0.0005 in. or more out of round.
5. When to renew main bearings is a judgment call, based on sound, feel, and the condition of the crankshaft. The factory has designed a clever screw-type tool for extracting and removing both bearings and crankshaft seals. Figure 7-54 illustrates the tool in the act of installing the magneto-side bearing. The snap ring was removed prior to bearing extraction and reinstalled to act as a stop during bearing assembly. The governor-bell crank must be removed for access to the PTO bearing. Renew the bell-crank seal after bearing installation.
   Caution: Main bearing and governor seals install with the flat (marked) sides out.

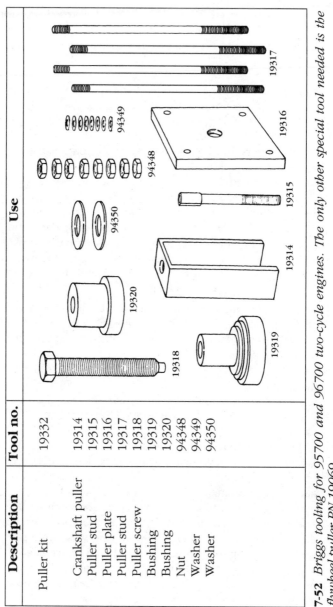

Description	Tool no.
Puller kit	19332
Crankshaft puller	19314
Puller stud	19315
Puller plate	19316
Puller stud	19317
Puller screw	19318
Bushing	19319
Bushing	19320
Nut	94348
Washer	94349
Washer	94350

7-52 *Briggs tooling for 95700 and 96700 two-cycle engines. The only other special tool needed is the flywheel puller PN 19069.*

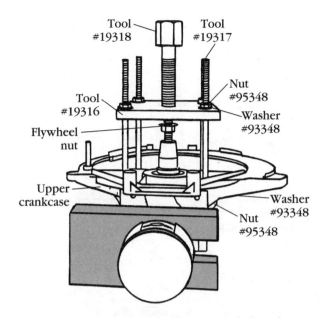

7-53 *The PTO-side crankcase half is removed, then the magneto half. In no case should the crankshaft be allowed to turn.* Briggs & Stratton Corp.

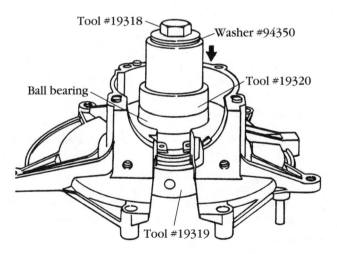

7-54 *Briggs main-bearing service tool provides local support for crankcase castings.*

6. Support the crankshaft in a vise (fitted with jaw protectors) and press the magneto-side crankcase casting home (Fig. 7-55) .

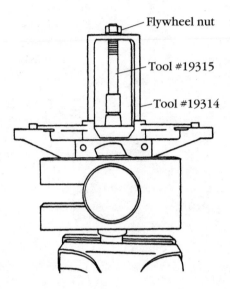

7-55 *The order of assembly is the reverse of disassembly. Position the magneto-side crankcase over the crankshaft and press the casting home.* Briggs & Stratton Corp.

Caution: Do not allow the crankshaft to turn during this and subsequent press operations.

7. Mount the flywheel on the crankshaft. Turn the partially completed assembly over to rest on the flywheel.
8. Position the crankcase gasket on the magneto-side casting flange. Orient the flange to index the locating pins and gasket holes.
9. Slide the governor assembly over the crankshaft with the weights tucked in (Fig. 7-56).
10. Rotate the governor-bell crank to the position shown on the PTO casting (Fig. 7-57).
11. Thread in one of the long puller studs (PN 19317 if you bought the kit) into the magneto-side casting to act as a pilot.

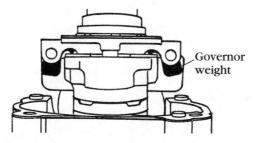

7-56 *Prior to assembly, retract governor weights as shown.* Briggs & Stratton Corp.

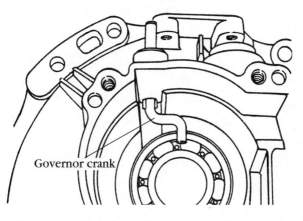

7-57 *The end of the bell crank lays on the bearing centerline.* Briggs & Stratton Corp.

12. Holding the governor-bell crank in position, press the crankshaft halves together.
    Caution: A sudden increase in the force level means that one or both bearings are misaligned in their bosses. Correct the problem before proceeding.
13. Progressively pull the screws down in an X-pattern to the 60 lb./in. torque limit.
14. Trim the gasket spillover (Fig. 7-58) with a razor blade and install the cylinder gasket and barrel. Torque cylinder screws as described in the previous step to 110 lb./in.

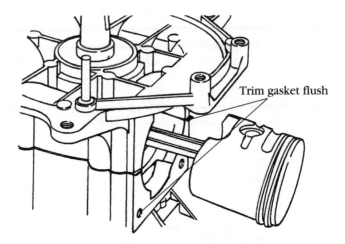

Trim gasket flush

7-58 *Trim surplus gasket material from the cylinder interface.*
Briggs & Stratton Corp.

8

The new Europa

Standard-series lawnmower engines come in various states of trim and performance. Quantum models have a long list of desirable features usually lacking on more plebeian models: automotive-type air cleaners, quiet mufflers, and centrifugal governors. The 3.75-hp Sprint is a higher performance version of the 3.5-hp 92900. Some otherwise very ordinary Standard engines can be ordered with cast-iron bores and heavier flywheels. But until recently, all Standard series (and most I/C series) engines were variations on the same formula and built with the same, or similar, tooling.

The 5-hp Europa 99700, introduced in 1992, represents a major shift in the company's engineering and marketing philosophies. While it retains certain family features associated with Briggs & Stratton products, the Europa is an entirely new design. "We had," said one of the engineers involved in the project, "a blank sheet of paper."

It's also the first Standard engine to cost more than a comparable I/C. Listing at $317, the Europa costs some $70 more than the 4-hp I/C 114900 and almost twice as much as the Standard 110900 version of that same engine. Prices are catalog list. A mower manufacturer would pay considerably less, but would still find the Europa an expensive proposition.

The name Europa refers to the shape of the shrouding, which supposedly reflects European design influence. Actually, the engine is an American product, designed in Milwaukee and built at the Burleigh plant in Wauwatosa, Wisconsin. This is worth noting because, not so long ago, it seemed possible that Briggs would devolve into a marketing operation, selling imports under its venerable name. V-twin engines are, in fact, built in Japan (with American tooling) and Mitsubishi makes the 4-, 5½-, and 9-hp Vanguard models.

Technical description

The Europa is only the first of a planned series of mechanically similar aluminum-block overhead-valve (OHV) engines. All will carry model numbers in the 99700/99799 range, and displace 147 cc from 64.5-mm bore and 44.2-mm stroke. Valves mount vertically above the piston and actuate through pushrods and stamped-steel rocker arms.

The crankshaft rides on block-metal bearings, a practice that Briggs engineers defend on the grounds that 90 percent of customers discard an engine before the main bearings fail. If necessary, the block can be reamed and fitted with bushings.

Two oiling systems are used, which calls to mind the old 216.5 CID Chevrolet that had scoopers on the connecting rods, squirters in the pan, and an oil pump. Splash provides the Europa with primary lubrication, vis-a-vis a paddle wheel driving off the camshaft gear. An eccentric oil pump, also camshaft-powered, oils the upper main bearing and rocker arms. A technician involved in Europa development said that the provision for upper main-bearing lubrication was inspired by Tecumseh practice.

As always, even in more expensive engines, the connecting rod runs directly against the crankshaft, without benefit of replaceable insert bearings. The crankpin can be reground to accommodate a 0.020-in. undersize rod.

A composite plastic and steel camshaft is used. Plastic cam lobes are a type of Briggs trademark. Although the camshafts work, there are trade-offs in terms of the rate of valve acceleration and allowable valve-spring forces. Europa valve springs are soft enough to be compressed with one finger.

Quality improves toward the top of the engine. Selectively fitted pistons, graded by size in increments of a hundredth (0.001) of a millimeter, run in thick iron liners that appear to have been centrifugally cast for grain uniformity. Combustion chambers receive nearly 100 percent machining to eliminate casting inaccuracies. Until recently, machined chambers were confined to diesel engines. Sintered iron bearings, which combine extreme hardness with porosity for oil retention, locate the rocker arms.

Because it is an entirely new design, the Europa required special tooling and remains expensive to produce. The number of parts was held to a minimum. Except for the valves themselves, all intake and exhaust components interchange. Bolt holes in the flange are numbered in the tightening torque sequence. A slight reduction in the diameter of the power-takeoff (PTO) end of the crankshaft allows the flange to be assembled without risk of damage to the lower oil seal.

Field experience

Europa engines are too new to have much of a track record and, without access to company warranty records, one is forced to depend on anecdotal information and public-relations claims. A company spokesman described the Europa and its spin-offs to come as the OEM-preferred choice, "which implies reliability." Snapper was the first mower manufacturer to make a large purchase. Engineering said they knew of no special problems with the design, which (aside from the valve gear) is quite conservative.

A check with Houston-area mechanics turned up several who professed ignorance of the engine and two who claimed to have witnessed or heard about catastrophic failure. The cause was dirt ingestion past dirty air filters. A counterman at a local Briggs distributor added an ingenious twist to the story: Other Briggs engines choke down when their filters clog, but the Europa keeps running until it self-destructs from inhaled grit.

The example studied for this book did, in fact, have an air cleaner problem that could have led to failure. It was corrected as described under the upcoming section "Assembly."

Overhead valves

The Europa is the first Standard-series engine to have its valves in the head over the piston. This configuration is by no means new. The Wright Flyer of 1903 was powered by an OHV engine. Chevrolet went to overhead valves in 1929, Ford and Chrysler in 1954. Indeed, the only side-valve engines that remain in series production are those made by Briggs and its arch-rival Tecumseh.

Overhead valves tidy up the combustion chamber, allowing it to be smaller and more symmetrical. With fewer places to hide, the air-fuel mixture burns more completely and with less tendency to detonate or knock. Consequently, compression ratio can be increased with gains in fuel economy and/or power. The Europa claims to develop 5 hp from a displacement of 8.99

cubic in. The side-valve Quantum 124700 requires almost 12 cubic in. to produce the same power.

Other advantages are more direct, generally less restrictive, porting and a cooler-running exhaust valve. The latter consideration is of major importance for an air-cooled engine.

The promise of higher output, lower maintenance, and a socially acceptable valve gear drove the design. The engine was already in production when control of exhaust emissions became the first priority of all small engine makers. For reasons explained in the following sections, emissions control begins with overhead valves.

Small-engine emissions

There was little corporate concern about exhaust emissions until 1993, when California and federal regulators woke up to the fact that small four-cycle engines are roughly 30 times dirtier than modern automobiles on a gram/hp basis. By another computation, one hour spent mowing the lawn generates about the same level of smog-producing emissions as driving a late-model car 80 miles. Two-cycle engines fare even worse. According to the California Air Resources Board, a chain saw generates as much emissions in two hours of hard use as an automobile does in 3,000 miles. Four types of emissions are considered critical:

- hydrocarbons (HC)
- nitrogen oxides (NOX)
- carbon monoxide (CO)
- carbon dioxide (CO_2).

HC emissions develop from unburned fuel and motor oil in the exhaust and percolate from the fuel tank, whether the engine is running or not. NOX forms during combustion, tending to become worse as flame temperatures increase. CO is present in the exhaust, but can be controlled by proper tuning. CO_2, the greenhouse gas, is a surcharge incurred for burning carbon-based fuels. The only way it can be limited at present is to consume less fuel.

Both the Environmental Protection Agency and California regulators have announced plans to crack down on small engines, starting in 1994. After initial resistance, the industry now admits it has a problem.

Short of some miracle in chemical engineering, the oil-burning two-cycle engine is doomed on the grounds of HC emissions. Surprisingly, these engines do fairly well in terms of NOX (exhaust gas from the previous cycle dilutes the incoming charge and quenches flame temperature).

The side-valve four-cycle is also a lost cause, primarily because of combustion-chamber geometry and secondly because no sane engineering department would invest the effort to clean up such an obsolete design. That leaves the OHV engine, although the Europa is not envisioned as an across-the-board replacement for Standard-series engines. The company would like to find a compromise design, clean enough to satisfy the EPA and inexpensive enough to compete at K-Mart. An F-head design (intake valve over the piston, exhaust alongside of it) is under investigation.

A closer look

I thought that it would be instructive to disassemble a Europa, especially since the factory has been slow to provide documentation. Obtaining an owner's manual and a parts list required a call to Milwaukee. After some delay, the local distributor was able to supply the specifications reprinted as Table 8-1. No shop manual existed as of late 1993.

Table 8-1.
Europa 99700–99799 specifications

General dimensions

Armature air gap	0.006"–0.014"
Spark-plug gap	0.030"
Standard cylinder bore	2.5625"–2.5615"
Valve clearance (cold, both)	0.003"–0.005"

Reject dimensions

Cam gear

Magneto journal	0.615"
PTO journal	0.615"

Crankshaft

Magneto journal	0.873"
Crankpin journal	1.122"
PTO journal	1.060"

Cylinder

Cylinder cam bearing	0.621"
Magneto main bearing	0.878"

Table 8-1. (Cont.)

Sump

| Sump cam bearing | 0.621" |
| Sump PTO main bearing | 1.065" |

Torque limits

Connecting rod screws	100 lb/in.
Cylinder head screws	160 lb/in.
Electric starter mounting screws	85 lb/in.
Flywheel nut	60 lb/in.
Governor lever lock nut	35–45 lb/in.
Sump screws	85 lb/in.

Daniel Faulkner, a Houston-based photographer specializing in auto racing and industrial work, shot the photographs that illustrate this section. For readers interested in the techniques of photographing machine parts, Mr. Faulkner used an F-4 Nikon with a Nikon 55-mm Micro-Nikkor lens, an Apollo soft box, and Norman flash equipment. Seamless white paper served as a backdrop.

Disassembly

Follow this procedure:

1. Some models come with a fuel cutoff valve. If your engine does not have this feature, drain the tank by disconnecting the hose at the carburetor fitting.
   Warning: Perform this operation outdoors in an area remote from potential ignition sources. As an additional precaution, place the control lever in the *stop* position.
2. Remove fuel tank assembly (integral with the decorative plastic shroud), oil-filler tube, and muffler guard.
   Note the plastic spacer between the carburetor-side, shroud-mounting lug and engine block.
3. Remove the three-piece blower housing and attached starter.
4. The combination of 60-lb./ft.-assembly torque on the flywheel nut and an aggressive taper fit means that the flywheel must be pulled, rather than knocked loose. Unless you have access to a factory puller, which uses self-tapping screws, it will be necessary to tap ¼ × 20 TP1 threads on the two holes cast into the hub, as shown in Fig. 8-1.

8-1 *Flywheel hub holes receive threads in preparation for disassembly.*

5. A suitable puller can be fabricated from bar stock with bolt holes on 1.5-in. centers (Fig. 8-2).
6. The Europa purchased for this exercise was a "compliance engine," meaning that it was fitted with a spring-loaded flywheel brake and ignition-shorting switch. The brake assembly can be unbolted or "cocked and locked" (Fig. 8-3).
7. Remove the air cleaner assembly, which is secured by three ⁵⁄₁₆-in. capscrews.
8. Use long-nosed pliers to unhook the throttle spring at the carburetor (Fig. 8-4).
9. Disengage the small spring that actuates the choke (Fig. 8-5). Although the spring is delicate and might rust, it is a major improvement over previous choke-engagement mechanisms.
10. Once the springs and ignition-kill wire are disconnected, gently rotate the carburetor assembly to release the governor link (Fig. 8-6).
11. Remove the valve cover, rocker-arm nuts, rocker arms, plastic valve spacers, and pushrods (Fig. 8-7).
   Caution: Set screws prevent rocker-arm balls from vibrating loose. These screws must be loosened before attempting to back off the balls.

8-2 *A homemade puller works about as well as the factory tool and costs nothing to fabricate.*

8-3 *A screwdriver blade locks the spring-loaded brake out of contact with the flywheel rim.*

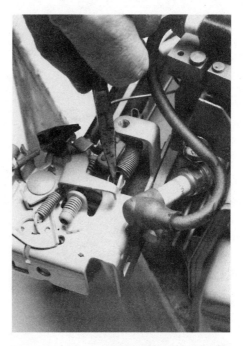

8-4 *Long-nosed pliers are used to disengage one end of the throttle spring.*

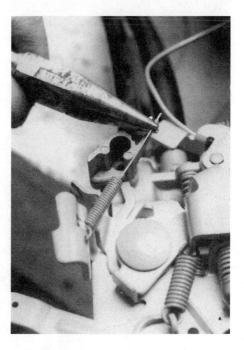

8-5 *The choke spring looks like the sort of part that will need frequent replacement. Order PN 262749.*

8-6 *The carburetor body rotates out of and into engagement with the governor link.*

8-7 *Partially disassembled cylinder head, showing the quality of foundry work now obtainable for a utility engine.*

8-8 *Four long bolts se-cure the cylinder head, which is stiffened by its compact imensions and large cooling fin area. Comparable side-valve heads require eight bolts and are by no means immune to gasket fail-ure. The arrow stamped on the piston crown is an assembly reference.*

12. Remove the four cylinder-head bolts (Fig. 8-8). While not mandatory, loosening these bolts in stages, working across the diagonals, reduces the risk of warping the head.
13. The oil pump is accessible from outside the flange (Fig. 8-9). Use a small screwdriver to pick out the impeller and rotor.
14. All new engines are contaminated to some degree with chips and drill dust. The example Europa was no exception. Most of the swarf was lodged at the oil-pump inlet, although some was distributed inside the crankcase. The factory apparently "motors" engines prior to shipment, rotating the crankshaft with an external power source.
    Note: Purchasers of these fairly expensive products would do well to clean the pump and crankcase before initial startup.
15. Using strip abrasive, thoroughly clean the crankshaft outside diameter (OD) in preparation for flange removal.
    Caution: Failure to thoroughly clean the crankshaft damages the oil seal and lower main bearing.
16. Undo the flange holddown cap screws and separate the flange from the block casting with a soft mallet. Carefully

8-9 *The oil pump drives off of the camshaft to lubricate the magneto-side main bearing and cylinder head components. Pressurized lurication should prolong the life of the upper main bearing, although this has not been proven. Tests of competitive engines with and without this feature show no measurable difference in bearing wear.*

withdraw the flange, which should slip easily off the crankshaft. The flyweight/oil-slinger assembly might drop into the flange (Fig. 8-10) but the camshaft should remain suspended in its magneto-side bearing to establish the original positions of the timing marks.

17. Check the timing as an assembly reference (Fig. 8-11). Unlike other engine makers, Briggs & Stratton does not index timing mark alignment with top dead center (tdc) on the compression stroke.

18. Withdraw the camshaft and tappets (Fig. 8-12). The camshaft assembly includes a compression release that holds the intake valve slightly open during cranking.

19. As always, become familiar with the rod-cap/shank orientation marks before proceeding with disassembly. The arrow on the cap points to the camshaft and aligns with a second arrow on the upper rod assembly (Fig. 8-13).

20. Withdraw the piston and upper rod assembly. Replaceable spring locks (circlips) locate the piston pin, which at room temperature should be a light interference fit with the piston. If further disassembly is required, warm the

8-10 *A camshaft-driven slinger provides primary lubrication in traditional Briggs fashion.*

8-11 *Mechanics soon learn to verify original valve timing before the camshaft is disturbed.*

8-12 *The Europa employs a composite plastic and steel camshaft that, like the fabled bumblebee, should not fly. In auto engines, camshaft-lobe pressures approach 50,000 psi at idle. Plastic lobes obviously work, however, as witnessed by the millions of Quantum engines that use them.*

8-13 *Connecting-rod match marks take the form of arrows pointing toward the camshaft. When oriented properly, the rod cap snaps onto the rod shank rabbet.*

piston to approximately 200 degrees Fahrenheit with a heat gun or place it (suitably protected) on an electric hot plate for a few seconds. The piston will expand enough to release the pin.

Caution: Driving the pin out cold can distort the piston.

Inspection

Clean parts with Varsol or an equivalent solvent. If a pressurized solvent gun is not available, submerge the block casting (less Magnetron and carburetor) in solvent for a few hours and blow out the oil passages with compressed air. Repeat until the discharge is clean.

Examine all friction surfaces for wear (Table 8-1) and surface finish. Lubricate with liberal amounts of motor oil (Fig. 8-14) or with engine-assembly lubricant. Fasteners should, if Briggs follows the earlier practice, be lightly oiled.

8-14 *In the absence of factory recommendations to the contrary, prudent mechanics should renew the connecting rod bolts upon disassembly. Torque to 100 lb./in.*

Obtain the replacement parts you need, which will include whatever gaskets have been disturbed, a spark plug, and air filter cartridge (PN 494586). Note that crankshaft-related parts vary somewhat between type numbers and that there was a running

parts change involving the cylinder-head casting and related
components. Existing documentation is not entirely clear on the
point but it appears that the change occurred on August 8, 1992,
or on build date 92080500. Because of differences in the head
gasket, two overhaul gasket sets are cataloged: PN 496055 after
92080500 and PN 494963 on and before 92080400.

Assembly

Follow this procedure:

1. According to Briggs engineering, worn main bearings can
   be reamed to accept bushings. Use the tooling developed
   for earlier engines and readily available to dealer
   mechanics.
2. Replace oil seals (magneto PN 299819, PTO PN 399781)
   as described in chapter 7.
3. Ideally, the crankshaft should be sent out to an
   automotive machinist for a thorough inspection and
   polishing At the minimum, mike bearing surfaces and
   look for bends outboard of the PTO bearing.
4. Replacement piston assemblies include ring sets, spring
   locks, and piston pins, which can also be purchased as
   individual items:

Size	Piston Assy.	Ring Set	Spring Lock	Pin
Standard	493781	493782	262514	262514
0.010 O/S	495267	495268	262514	262514
0.020 O/S	495269	495270	262514	262514
0.030 O/S	496271	496272	262514	262514

If the piston pin has been disturbed, heat the piston
as described in #20 under the previous section
"Disassembly." Using new spring locks, assemble the
piston, pin, and rod. Verify piston-to-rod orientation. If
incorrect, the engine will knock.

   Note: Connecting-rod assemblies are available in
standard (PN 493049) and 0.020-in. undersizes
(PN 493689) for use with reground crankshafts.

   Caution: Without factory documentation, the wisest
course of action is to renew the rod bolts (PN 94404)
upon assembly.

5. Using a pump-type dispenser, flood the rod bearings and
   crankshaft journal with oil. Verify that both connecting-
   rod arrows point toward the camshaft and assemble the

cap to the upper rod. Listen for a click when the cap seats.

6. Lightly lubricate new rod bolts with motor oil and torque down evenly in three increments to 100 lb./in. (Fig. 8-14). Turn the crankshaft through two full revolutions to detect possible binding.

7. Insert valve tappets into their respective bores. (Tappets interchange but should be returned to their original places on used engines.) Some mechanics apply assembly grease to the tappets and cam lobes as insurance against scuffing. Grease also prevents the tappets from falling out as the camshaft is installed.

8. Inspect the crankshaft key (PN 94388) for damage. Install the timing gear over the end of the crankshaft with the timing mark visible.

9. Install the camshaft, turning the crankshaft as necessary to index timing marks.

10. The slinger/governor hangs off the camshaft stub with the paddle-shaped governor lever resting on it and positioned over the recess cast into the floor of the flange (Fig. 8-15).

8-15 *Critical wear sites include the camshaft lobes (replace if pitted or galled), camshaft journals, com-pression release, and cam-shaft gear teeth. Tappets should be returned to their original bores and liberally coated with assembly lube.*

11. Position a new flange gasket on the crankcase casting. Because the factory supplies only one gasket (PN 2721324), crankshaft endplay can be ignored. A thinner gasket could be used on earlier engines to compensate for thrust-bearing wear.

12. Mount the flange over the crankshaft stub, pressing it home with the palm of your hand.

13. Progressively torque flange screws in the sequence indicated on the casting to the 85 lb./in. torque limit (Fig. 8-16).

8-16 *Flange screws should be lightly oiled and torqued in the indicated sequence to 85 lb./in.*

14. Oil pump wear limits have not yet been published, but new engines appear to be set up with 0.002 in. between the rotor OD and recess (Fig. 8-17). Because of its location at the bottom of the sump, the pump cover appears vulnerable to abrasion. The pump rotor and drive gear are listed under PN 493884. The cover and O-ring under PN 4927656.

15. The cylinder-head gasket (currently PN 272314; PN 272488 before 92080500) mounts either side up. Head bolts should be lightly oiled and torqued in an X-pattern to 160 lb./in.

8-17 *Most pump wear should occur on the cover, which can be replaced inexpensively.*

16. Experienced mechanics judge valve guide condition by eye: a bell-mouthed guide allows the valve to wobble in the fully open position. More precise methods await the release of factory documentation, which should also detail the required tooling and extraction/installation procedure. In the interim, entrust valve guide work to a competent automotive machinist, who should also be able to approximate factory valve-face angles.

17. Install the valves, noting that the intake valve assembly includes an umbrella-type oil seal (PN 493661) and seal gasket (PN 272376). No special tools are needed because the interchangeable valve springs are soft enough to compress by hand.

18. Install the rocker assemblies in this sequence: pushrod guide plate (marked *top*), pushrods, plastic valve caps (PN 262499), rocker arms, and ball nuts. Apply assembly lube to the rocker-arm/ball-nut contact surfaces.

19. Rotate the crankshaft as necessary to bring the tappets on the heel of the camshaft lobes and, on the intake side, out of engagement with the compression release. Set

valve lash to the 0.003- to 0.005-in. cold specification (Fig. 8-18). Tighten the ball-nut set screws to fix the adjustment.

20. Install the flywheel key (PN 222698), flywheel, starter hub, and flywheel nut. Torque the nut to 60 lb./ft.
21. Complete the assembly by installing the carburetor, muffler, and miscellaneous parts.
22. The air cleaner assembly requires special mention. It consists of a foam precleaner and a paper cartridge, sealed by pliable gaskets and contained in a plastic housing. A careful inspection prior to disassembly revealed no obvious leak paths between the air cleaner and the carburetor intake.

Upon installation, an ⅛-in. gap remained between the bottom of the carburetor air horn and the air cleaner mounting surface—even with the hold-down bolts torqued to the shear point. The carburetor flange threads had been cut with a tap worn undersized. Chasing the threads with a 10-32 tap corrected the problem (Fig. 8-19).

8-18 *Static valve adjustments are made on a cold engine with rocker-ball set screws backed off and tappets on the cam-base circle. Make certain the intake tappet rests in its fully retracted position, clear of the compression release. Tightening the rocker-ball set screws against their respective studs fixes the adjustment.*

8-19 *It was necessary to retap the air cleaner-to-carburetor-screw holes for the example engine.*

How many Europas came off the line with impossible-to-retighten air filter screws is anyone's guess. Reported tales of catastrophic failure associated with dirty air cleaners might have a basis in fact. It is also true that a leaking Europa will run to self-destruction with a clogged air filter.

Index

215

About the author

Paul Dempsey is the Senior Consultant for Engineering Management Consultants, a Houston-based consulting firm, and the author of more than 20 technical books. He is a member of Instrument Society of America, International Maintenance Institute, American Petroleum Institute, Automotive Engine Rebuilders Association, International Association of Drilling Contractors, and a standing committee member for the Offshore Marine Services Association.

Dempsey is the author of more than 20 TAB books, two of which sold more than 100,000 copies. Dempsey's broad familiarity with mechanical devices combined with his skill and knowledge of writing technical manuscripts, has prompted him to write more than 100 magazine and journal articles dealing with English literature, teaching techniques, petroleum-related subjects, and maintenance management.

Other Bestsellers of Related Interest

**Troubleshooting and
Repairing Diesel Engines, 3rd Edition**
—Paul Dempsey
This easy-to-read handbook gives owners and mechanics guidance on how to maintain, ttroubleshoot, and repair the latest diesel engines in farm and lawn equipment, boats, cars, trucks, air compressors, and generators.
   ISBN 0-07-016348-0 $24.95 Paper

Small Gas Engine Repair, 2nd Edition
—Paul Dempsey
A fast and easy introduction to the fascinating world of international shortwave radio listening. Covers history, equipment, terminology, station profiles, broadcast schedules, resources, and much more.
   ISBN 0-8306-4142-4 $11.95 Paper
   ISBN 0-8306-4141-6 $24.95 Hard

How to Order

☎ **Call 1-800-822-8158**
24 hours a day,
7 days a week
in U.S. and Canada

✉ **Mail this coupon to:**
McGraw-Hill, Inc.
Blue Ridge Summit, PA
17294-0840

📠 **Fax your order to:**
717-794-5291

💻 **EMAIL**
70007.1531@COMPUSERVE.COM
COMPUSERVE: GO MH

Thank you for your order!

Shipping and Handling Charges

Order Amount	Within U.S.	Outside U.S.
Less than $15	$3.45	$5.25
$15.00 - $24.99	$3.95	$5.95
$25.00 - $49.99	$4.95	$6.95
$50.00 - and up	$5.95	$7.95

EASY ORDER FORM—
SATISFACTION GUARANTEED

Ship to:

Name _____

Address _____

City/State/Zip _____

Daytime Telephone No. _____

ITEM NO.	QUANTITY	AMT.

Method of Payment:

☐ Check or money order
enclosed (payable to
McGraw-Hill)

☐ [card]   ☐ *VISA*

☐ MasterCard   ☐ DISCOVER

Shipping & Handling charge from chart below	
Subtotal	
Please add applicable state & local sales tax	
TOTAL	

Account No. [ ][ ][ ][ ][ ][ ][ ][ ][ ][ ][ ][ ][ ]

Signature _____ Exp. Date _____
Order invalid without signature

**In a hurry? Call 1-800-822-8158 anytime,
day or night, or visit your local bookstore.**

Code = BC44ZNA